AF352859

# The 1st International Online Conference on Infrastructures

# The 1st International Online Conference on Infrastructures

Editors

**Joaquín Martínez-Sánchez**
**Patricija Kara De Maeijer**
**Davide Lo Presti**
**Hosin "David" Lee**
**Ana Sánchez Rodríguez**
**Francesco Liberati**

MDPI • Basel • Beijing • Wuhan • Barcelona • Belgrade • Manchester • Tokyo • Cluj • Tianjin

*Editors*

Joaquín Martínez-Sánchez
Escuela de Ingeniería de
Minas y Energía
Spain

Patricija Kara De Maeijer
University of Antwerp
Belgium

Davide Lo Presti
University of Palermo
Italy

Hosin "David" Lee
University of Iowa
USA

Ana Sánchez Rodríguez
Universidade de Vigo
Spain

Francesco Liberati
Sapienza University of Rome
Italy


*Editorial Office*
MDPI
St. Alban-Anlage 66
4052 Basel, Switzerland


This is a reprint of articles from the Proceedings published online in the open access journal *Engineering Proceedings* (ISSN 2673-4591) (available at: https://www.mdpi.com/2673-4591/17/1).

For citation purposes, cite each article independently as indicated on the article page online and as indicated below:

LastName, A.A.; LastName, B.B.; LastName, C.C. Article Title. *Journal Name* **Year**, *Volume Number*, Page Range.

ISBN 978-3-0365-4963-7 (Hbk)
ISBN 978-3-0365-4964-4 (PDF)

# Contents


**About the Editors** . . . . . . . . . . . . . . . . . . . . . . . . . . . . . . . . . . . . . . . . . . . . . . . . . . **ix**

**Preface to "The 1st International Online Conference on Infrastructures"** . . . . . . . . . . . . . . **xi**

**S. Sonny Kim**
Assessment of Pavement Structural Conditions Using a Ground-Penetrating Radar
Reprinted from: *Eng. Proc.* **2021**, *17*, 1, doi:10.3390/engproc2022017001 . . . . . . . . . . . . . . . **1**

**Ahmed Al-Mohammedawi and Konrad Mollenhauer**
Filler Effect on Moisture Resistance of Cold Recycling Materials
Reprinted from: *Eng. Proc.* **2021**, *17*, 2, doi:10.3390/engproc2022017002 . . . . . . . . . . . . . . . **3**

**Alejandra Ospina-Bohórquez, Jorge López-Rebollo, Pedro Muñoz-Sánchez and
Diego González-Aguilera**
A Digital Twin for Monitoring the Construction of a Wind Farm
Reprinted from: *Eng. Proc.* **2021**, *17*, 3, doi:10.3390/engproc2022017003 . . . . . . . . . . . . . . . **7**

**Gurkan Yildirim, Ashraf Ashour, Emircan Ozcelikci, Muhammed Faruk Gunal and
Behlul Furkan Ozel**
Development of Geopolymer Binders with Mixed Construction and DemolitionWaste-Based
Materials
Reprinted from: *Eng. Proc.* **2021**, *17*, 4, doi:10.3390/engproc2022017004 . . . . . . . . . . . . . . . **9**

**Linh Truong-Hong**
Transform Physical Assets to 3D Digital Models
Reprinted from: *Eng. Proc.* **2021**, *17*, 5, doi:10.3390/engproc2022017005 . . . . . . . . . . . . . . . **11**

**Didier Snoeck**
Autogenous Healing in 10-Years Aged Cementitious Samples Containing Microfibers and
Superabsorbent Polymers
Reprinted from: *Eng. Proc.* **2021**, *17*, 6, doi:10.3390/engproc2022017006 . . . . . . . . . . . . . . . **13**

**Sebastian Esser**
Graph-Based Version Control of BIM Models in an Event-Driven Collaboration Environment
Reprinted from: *Eng. Proc.* **2021**, *17*, 7, doi:10.3390/engproc2022017007 . . . . . . . . . . . . . . . **17**

**Wee Teo, Kazutaka Shirai and Jee Hock Lim**
Influence of Precursor Materials on the Mechanical Behavior of Ambient-Cured One-Part
Engineered Geopolymer Composites
Reprinted from: *Eng. Proc.* **2021**, *17*, 8, doi:10.3390/engproc2022017008 . . . . . . . . . . . . . . . **19**

**Edoardo Bocci, Emiliano Prosperi and Maurizio Bocci**
Evaluation of Binder-Aggregate Adhesion in Hot-Recycled Asphalt Mixtures as a Function of
the Production Temperature
Reprinted from: *Eng. Proc.* **2021**, *17*, 9, doi:10.3390/engproc2022017009 . . . . . . . . . . . . . . . **21**

**Vincent P. Pilien, Lessandro Estelito O. Garciano, Michael Angelo B. Promentilla,
Ernesto J. Guades, Julius L. Leaño, Andres Winston C. Oreta and
Jason Maximino C. Ongpeng**
Banana Fiber-Reinforced Geopolymer-Based Textile-Reinforced Mortar
Reprinted from: *Eng. Proc.* **2021**, *17*, 10, doi:10.3390/engproc2022017010 . . . . . . . . . . . . . **23**

**Marianna Loli, John Manousakis, Stergios A. Mitoulis and Dimitrios Zekkos**
UAVs for Disaster Response: Rapid Damage Assessment and Monitoring of Bridge Recovery
after a Major Flood
Reprinted from: *Eng. Proc.* **2021**, *17*, 11, doi:10.3390/engproc2022017011 . . . . . . . . . . . . . **25**

**Vyacheslav Falikman**
Defined-Performance Concretes Using Nanomaterials and Nanotechnologies
Reprinted from: *Eng. Proc.* **2021**, *17*, 12, doi:10.3390/engproc2022017012 . . . . . . . . . . . . . **27**

**Muhammad Imran and Francesco Liberati**
A Framework for Intelligent Decision Making in Networks of Heterogeneous Systems (UAV
and Ground Robots) for Civil Applications
Reprinted from: *Eng. Proc.* **2021**, *17*, 13, doi:10.3390/engproc2022017013 . . . . . . . . . . . . . **29**

**Daniel Grossegger**
Resource Efficiency to Achieve a Circular Economy in the Asphalt Road Construction Sector
Reprinted from: *Eng. Proc.* **2021**, *17*, 14, doi:10.3390/engproc2022017014 . . . . . . . . . . . . . **31**

**Byungkyu Moon and Hosin (David) Lee**
Drone-Image Based Fast Crack Analysis Algorithm Using Machine Learning for Highway
Pavements
Reprinted from: *Eng. Proc.* **2021**, *17*, 15, doi:10.3390/engproc2022017015 . . . . . . . . . . . . . **33**

**Maria Pina Limongelli, Carmelo Gentile, Fabio Biondini, Marco di Prisco, Francesco Ballio
and Marco Belloli**
The Monitoring Guidelines of the Lombardia Region in Italy
Reprinted from: *Eng. Proc.* **2021**, *17*, 16, doi:10.3390/engproc2022017016 . . . . . . . . . . . . . **37**

**Salvatore Antonio Biancardo, Cristina Oreto, Rosa Veropalumbo and Francesca Russo**
Integrated BIM-Based LCA for Road Asphalt Pavements
Reprinted from: *Eng. Proc.* **2021**, *17*, 17, doi:10.3390/engproc2022017017 . . . . . . . . . . . . . **39**

**Orazio Baglieri, Anna Viola, Arianna Fonsati and Anna Osello**
Pavement Information Modelling (PIM): Best Practice to Build a Digital Repository for Roads
Asset Management
Reprinted from: *Eng. Proc.* **2021**, *17*, 18, doi:10.3390/engproc2022017018 . . . . . . . . . . . . . **41**

**Christian Paglia, Giuliano Frigeri and Armando Chollet**
The Use of Tunnel Demolition Rocks to Produce Shotcrete for a Railway Infrastructure
Reprinted from: *Eng. Proc.* **2021**, *17*, 19, doi:10.3390/engproc2022017019 . . . . . . . . . . . . . **43**

**Gaetano Di Mino, Konstantinos Mantalovas and Vineesh Vijayan**
Investigating the Viability of Multi-Recycling of Asphalt Mixtures through a Preliminary Binder
Level Characterization
Reprinted from: *Eng. Proc.* **2021**, *17*, 20, doi:10.3390/engproc2022017020 . . . . . . . . . . . . . **45**

**Andrea Marcucci, Stefano Guanziroli, Alberto Negrini, Liberato Ferrara and Bernardino
Chiaia**
Introduction to a New Extrusion-Based Technology for the Regeneration of Existing Tunnels
Reprinted from: *Eng. Proc.* **2021**, *17*, 21, doi:10.3390/engproc2022017021 . . . . . . . . . . . . . **47**

**Sandhya Makineni, Avishreshth Singh and Krishna Prapoorna Biligiri**
Lifecycle Assessment of Permeable Interlocking Concrete Pavement and Comparison with
Conventional Mixes
Reprinted from: *Eng. Proc.* **2021**, *17*, 22, doi:10.3390/engproc2022017022 . . . . . . . . . . . . . **49**

**L. M. González-deSantos**
News Applications of UAVs for Infrastructure Monitoring: Contact Inspection Systems
Reprinted from: *Eng. Proc.* **2021**, *17*, 23, doi:10.3390/engproc2022017023 . . . . . . . . . . . . . . **51**

**Kruthi Kiran Ramagiri, Patricia Kara De Maeijer and Arkamitra Kar**
High-Temperature, Bond, and Environmental Impact Assessment of Alkali-Activated Concrete
(AAC)
Reprinted from: *Eng. Proc.* **2021**, *17*, 24, doi:10.3390/engproc2022017024 . . . . . . . . . . . . . . **53**

**Iain Peter Dunn, Francesco Acuto, Oumaya Yazoghli-Marzouk, Gaetano Di Mino and
Konstantinos Mantalovas**
Increasing the Use of Reclaimed Asphalt in Italy towards a Circular Economy: A Top-Down
Approach
Reprinted from: *Eng. Proc.* **2021**, *17*, 25, doi:10.3390/engproc2022017025 . . . . . . . . . . . . . . **55**

**Thien Nhan Tran, Salvatore Mangiafico, Cédric Sauzéat and Hervé Di Benedetto**
Bituminous Interlayers Thermomechanical Behaviour under Small Shear Strain Loading Cycles
with 2T3C Apparatus: Hollow Cylinder and Digital Image Correlation
Reprinted from: *Eng. Proc.* **2021**, *17*, 26, doi:10.3390/engproc2022017026 . . . . . . . . . . . . . . **57**

**Onur Ozturk, Hasan Yildirim, Nilufer Ozyurt and Turan Ozturan**
Mechanical Properties and Structural Requirements of Recycled Aggregate Concrete for
Pavements
Reprinted from: *Eng. Proc.* **2021**, *17*, 27, doi:10.3390/engproc2022017027 . . . . . . . . . . . . . . **61**

**Juliana O. Costa, Flavio A. dos Santos, Augusto C. S. Bezerra, Paulo H. R. Borges, Johan Blom
and Wim Van den bergh**
Reclaimed Asphalt and Alkali-Activated Slag Systems: The Effect of Metakaolin
Reprinted from: *Eng. Proc.* **2021**, *17*, 28, doi:10.3390/engproc2022017028 . . . . . . . . . . . . . . **63**

**Rassidatou Abbo, Okpwe Mbarga Richard, Lezin Seba Minsili and Mbondo Jean Marc**
A Bim Approach for the Design of a 5D Model of Industrial Warehouses in the Marine
Environment
Reprinted from: *Eng. Proc.* **2021**, *17*, 29, doi:10.3390/engproc2022017029 . . . . . . . . . . . . . . **67**

**Neetu G. Kumar, Avishreshth Singh and Krishna Prapoorna Biligiri**
Numerical Simulation of Pavement Subbase LayerModified with Recycled Concrete Aggregates
and Tire Derived Aggregates
Reprinted from: *Eng. Proc.* **2021**, *17*, 30, doi:/10.3390/engproc2022017030 . . . . . . . . . . . . . **69**

**Lakshmi Thotakura, Ganesh Babu Kodeboyina, Deepti Avirneni and Sankar Kumar Reddy
Pullalacheruvu**
Sustainability of Infrastructure and the Need for a Reassessment
Reprinted from: *Eng. Proc.* **2021**, *17*, 31, doi:10.3390/engproc2022017031 . . . . . . . . . . . . . . **71**

**Onur Ozturk and Nilufer Ozyurt**
Effects of Polypropylene Macro Fibers on the Structural Requirements, Cost and Environmental
Impact of Concrete Pavements
Reprinted from: *Eng. Proc.* **2021**, *17*, 32, doi:10.3390/engproc2022017032 . . . . . . . . . . . . . . **73**

**Štefan Jaud and Sergej Muhič**
Checking IFC with MVD Rules in Infrastructure: A Case Study
Reprinted from: *Eng. Proc.* **2021**, *17*, 33, doi:10.3390/engproc2022017033 . . . . . . . . . . . . . . **75**

**Gabriella Buttitta, Gaspare Giancontieri, Silvia Milazzo, Chiara Mignini,**
**Patricia Hennig Osmari, Usman Ghani and Davide Lo Presti**
Investigating Tools for Sustainability Assessment of Road Pavements in Europe
Reprinted from: *Eng. Proc.* **2021**, *17*, 34, doi:10.3390/engproc2022017034 . . . . . . . . . . . . . . **77**

**Ana Jiménez del Barco Carrión**
Combining Reclaimed Asphalt and Non-Petroleum-Based Binders for the Design of Sustainable
Asphalt Mixtures
Reprinted from: *Eng. Proc.* **2021**, *17*, 35, doi:10.3390/engproc2022017035 . . . . . . . . . . . . . . **81**

**Guang Ye**
Alkali-Activated Materials as Alternative Binders for Structural Concrete: Opportunities and
Challenges
Reprinted from: *Eng. Proc.* **2021**, *17*, 36, doi:10.3390/engproc2022017036 . . . . . . . . . . . . . . **83**

**John Harvey**
Improving Pavement Sustainability through Integrated Design, Construction, Asset
Management, LCA and LCCA
Reprinted from: *Eng. Proc.* **2021**, *17*, 37, doi:10.3390/engproc2022017037 . . . . . . . . . . . . . . **85**

# About the Editors

**Joaquín Martínez-Sánchez**

Dr. Joaquín Martínez-Sánchez, PhD and Assistant Professor at the University of Vigo. He has co-authored more than 80 papers and participated in several European and National R&D projects. His main research interests are Geomatic techniques applied to Transport infrastructure, integration of Sensors in Mobile Mapping Systems and data processing and analysis methods.

**Patricia Kara De Maeijer**

Patricia Kara De Maeijer is a civil engineer and carries out research in EMIB research group at UAntwerp (Belgium). Patricia's main research areas are concrete and alkali-activated materials technology, recycling of industrial wastes and by-products, asphalt and bitumen, and fiber Bragg grating (FBG) sensors monitoring systems. Patricia is a Senior member of RILEM. Patricia has (co-)authored more than 60 scientific publications. Patricia acts as a peer-reviewer in several scientific journals and recently has edited a book "Recent Advances and Future Trends in Pavement Engineering" in Infrastructures MDPI.

**Davide Lo Presti**

Davide Lo Presti is an Assistant Professor and holder of the "Rita Levi Montalcini fellowship" at the University of Palermo, Italy, since 2019, and a Visiting Academic at the University of Nottingham (UK). He graduated from the University of Palermo, Faculty of Engineering in 2005, and obtained his master's thesis in 2007. He defended his PhD thesis in collaboration with the University of Nottingham in 2011. He held several research positions at the University of Nottingham, Nottingham Transportation Engineering Centre, from 2011 to 2019. He worked as a Visiting Researcher at TU Delft, University of Washington, University of Sao Paolo and University of California, Davis.

**Hosin "David" Lee**

Dr. Hosin "David" Lee is a Professor in Civil and Environmental Engineering and a Director of Laboratory for Advanced Construction Technology (LACT), Iowa Technology Institute, University of Iowa. Dr. Lee is an internationally recognized expert in pavement engineering and infrastructure asset management. Dr. Lee currently serves Associate Editor of MDPI Infrastructures Journal and Editorial Board of ASCE Journal of Infrastructure System. He is a Member of National Academy of Engineering of Korea. He served Presidents of iSMARTi, KSEA, KSCEE and KOTAA, Board of Trustees of Seoul Institute of Technology, and Chairman of ASCE Highway Pavements Committee. He received the awards from Ministry of Science and ICT of Korea, Asphalt Paving Association of Iowa (APAI), Utah Engineers Council, iSMARTi, KSEA, KOFST, KSRE, and ASCE-Utah Section. He received the Otto Monsted Professorship and COWI Foundation Fellowships in Denmark. He served as Chairman of the First International Conference on Smart Cities, First International Conference on Maintenance and Rehabilitation of Constructed Infrastructure Facilities, 24th US-Korea Conference, and Fifth International Conference on Maintenance and Rehabilitation of Pavements.

**Ana Sánchez Rodríguez**

Ana Sánchez Rodríguez is a postdoctoral researcher at University of Valencia (Spain). Her current research is focused on automatic processing of LiDAR point clouds for inventory and inspection purposes, as well as the creation of digital models of infrastructure assets. Her research lines include the inspection and monitoring of bridges to avoid failure propagation.

**Francesco Liberati**

Francesco Liberati is a researcher in Automatic Control at the Sapienza University of Rome. He currently works mainly on cyber-physical systems and control problems in smart grids. He obtained his PhD in Systems Engineering from Sapienza University, with a dissertation over recent control problems in the area of energy management in smart grids. From May 2017 to December 2018, he was H2020 energy research project manager at "The Innovation and Networks Executive Agency (INEA)", European Commission, Brussels, where he managed a portfolio of large energy and smart cities H2020 research projects. Previously, he carried out applied research in several European funded projects, also with project management roles as team leader, work package leader, task leader.

**Mario Soilán**

Mario Soilán is a postdoctoral researcher at University of Salamanca (Spain). His current research is focused on the automatization of processes for the digitalization of infrastructure using geomatic information such as 3D point clouds, as well as on the development of geomatic data processing algorithms for road maintenance and inventory.

# Preface to "The 1st International Online Conference on Infrastructures"

The long-term preservation of Transport Infrastructure is a multidisciplinary effort. Transport Infrastructure (TI) is a key and strategic asset that supports social and economic growth and development through communication and mobility of people and goods. At the same time, roads are also essential for sustainability, the development of resilient transportation networks being one of the Sustainable Development Goals (SDG) for the UN's 2030 Agenda.

There is a need for new visions to improve the efficiency and resilience of infrastructures and their availability in safe and secure conditions for a better mobility.

In order to achieve this goal, prioritizing investments is key, both for new projects and the maintenance of existing infrastructure. This goal has been addressed through different research programmes at national and international levels, considering several aspects that can be simplified in KPIs such as in the acronym RAMSSHE€P, which stands for Reliability, Availability, Maintainability, Safety, Security, Health, Environment, €conomics, and Politics.

These societal, maintenance, and performance aspects configure an information vector of impacts, which defines a challenging constraint for predictive maintenance, providing as much information as possible to infer the condition rating (CR) of the TI or asset and thus prevent critical problems. The main idea is to forecast asset deterioration to support infrastructure management and to combine quantitative and objective collection of data with a processing and analysis tool chain during the whole lifecycle of the infrastructure.

Moreover, in the current context of the COVID-19 pandemic and its significant impact on the supply chain, demand for raw materials and transportation costs have increased, resulting in shortages and price hikes. As a result, construction works are more expensive. This gives even more importance to the need for a focus on Sustainability, fostering recycling and the reuse of materials, one of the essential aims of a Circular Economy Action Plan.

Construction and demolition (C&D) waste, including all the waste generated by the construction and demolition of buildings and infrastructures, as well as road maintenance, is an appreciable share of all the waste produced in the world. Due to the volume of materials involved in this process, there is a need for improving circularity because of the large potential of these valuable resources. This implies both managing C&D waste in an environmentally sound way and contributing to the transition to a circular economy through valorisation in high-value applications.

Apart from the Green Transition, there is a need for a digital transition in infrastructure management. BIM is an information model that represents the physical and attribute elements of construction and permits the sharing of knowledge and practices in a collaborative environment throughout the building lifecycle.

Exploring this information model can be powerful for high-value applications, from the design all the way to the final demolition. Infrastructure managers can find this high value from accurate and multisource data inputs that are updated periodically through infrastructure monitoring to collaboratively forecast the risks and challenges for all the involved stakeholders, with emphasis on collaboration. This is a key to improve a service that is essential for people and economy: mobility.

**Joaquín Martínez-Sánchez , Patricija Kara De Maeijer, Davide Lo Presti, Hosin "David" Lee ,
Ana Sánchez Rodríguez, and Francesco Liberati**

*Editors*

*Abstract*

# Assessment of Pavement Structural Conditions Using a Ground-Penetrating Radar [†]

S. Sonny Kim

School of Environmental, Civil, Agricultural, and Mechanical Engineering, College of Engineering, The University of Georgia, Athens, GA 30602, USA; kims@uga.edu; Tel.: +1-706-542-9804
† Presented at the 1st International Online Conference on Infrastructures, 7–9 June 2022; Available online: https://ioci2022.sciforum.net/.

**Keywords:** ground-penetrating radar (GPR); pavement structure; electromagnetic mixing theory; subgrade soil density



**Citation:** Kim, S.S. Assessment of Pavement Structural Conditions Using a Ground-Penetrating Radar. *Eng. Proc.* **2022**, *17*, 1. https://doi.org/10.3390/engproc2022017001

Academic Editor: Joaquín Martínez-Sánchez

Published: 10 May 2022

**Publisher's Note:** MDPI stays neutral with regard to jurisdictional claims in published maps and institutional affiliations.



Ground-penetrating radar (GPR) technology has been widely applied in ground subsurface investigations. Since 1980s, GPR technology has become a well-established technique for pavement performance evaluation. The analysis of GPR data provides rich information on the layer depths of pavement structures, material conditions, moisture content, voids, and locations of reinforcement and other features. The capability to accurately and reliably assess the subsurface conditions of pavement structure is essential to investigate both functional and structural deficiencies of pavements and their associated causalities resulting in the most cost-effective maintenance and rehabilitation treatments. The overall goal of this study is to extend GPR technology in combination with modern data analytics to provide improved pavement investigation methodology. In this study, an analytical approach to estimate field subgrade density is presented, which is critical for the diagnosis of pavement foundation failure [1,2].

**Funding:** The work presented in this paper is part of a research project (RP 19–21) sponsored by the Georgia Department of Transportation (GDOT). The contents of this paper reflect the views of the authors, who are solely responsible for the facts and accuracy of the data, opinions, and conclusions presented herein. The contents may not reflect the views of the funding agency or other individuals.

**Institutional Review Board Statement:** Not applicable.

**Informed Consent Statement:** Not applicable.

**Data Availability Statement:** Not applicable.

**Acknowledgments:** The authors would like to thank the many GDOT personnel that assisted with this study. A special thanks to Ian Rish, P.E., GDOT State Pavement Engineer.

**Conflicts of Interest:** The author declares no conflict of interest. The funding sponsors had no role in the design of the study; in the collection, analyses, or interpretation of data; in the writing of the manuscript, and in the decision to publish the results.

## References

1. Abdelmawla, A.M.; Durham, S.A.; Kim, S. Estimation of Subgrade Soil Density Using Ground Penetrating Radar. In Proceedings of the 2019 ICSC Conference, Seoul, Korea, 17–19 July 2019.
2. Abdelmawla, A.; Kim, S. Application of Ground Penetrating Radar to Estimate Subgrade Soil Density. *Infrastructures* **2020**, *5*, 12. [CrossRef]

MDPI

*Abstract*

# Filler Effect on Moisture Resistance of Cold Recycling Materials †

**Ahmed Al-Mohammedawi *** and **Konrad Mollenhauer**

Engineering and Maintenance of Road Infrastructure, Transportation Institute, University of Kassel, Mönchebergstraße 7, 34125 Kassel, Germany; k.mollenhauer@uni-kassel.de
* Correspondence: a.al-mohammedawi@uni-kassel.de

† Presented at the 1st International Online Conference on Infrastructures, 7–9 June 2022;
   Available online: https://ioci2022.sciforum.net/.

**Keywords:** moisture resistance; bitumen emulsion; active filler

**Citation:** Al-Mohammedawi, A.;
Mollenhauer, K. Filler Effect on
Moisture Resistance of Cold
Recycling Materials. *Eng. Proc.* **2022**,
*17*, 2. https://doi.org/10.3390/
engproc2022017002

Academic Editor: Andrea Grilli

Published: 2 May 2022

**Publisher's Note:** MDPI stays neutral
with regard to jurisdictional claims in
published maps and institutional affiliations.



## 1. Overview and Novelty

Cold recycling materials (CRM) with bitumen emulsion are getting increasingly important, aiming at highly efficient road infrastructure and tackling energy consumption, as well as its further consequences on climate change. Normally, cement is added to get improved strength, but its usage leads to risk, again, in mixture performance, such as brittleness behavior and drying shrinkage [1,2]. The objective of the present study is to analyze how eco-friendly by-product fillers affect the moisture resistance, as well as the stiffness of CRM.

## 2. Methodology and Results

The aggregate blend of the mortars was obtained by removing the coarse aggregate (larger than 2 mm) CRM granulate. The emulsion and filler content was fixed to 5% emulsion content and filler content of 3%. Cationic slow-setting bitumen emulsion was used. Various fillers were selected to provide an extensive overview of the effect of fillers on the mechanical properties and water sensitivity of CRM materials: cement (CE), ladle slag (LD), silica fume (SF), Ettringite binder (ET:70% LD + 30% gypsum), and geopolymer (GO:55% LD 35% Fly ash + 10% SF). Two different methods were used to assess the water sensitivity which are Rolling Bottle Test (RBT) and Shaking Abrasion Test (SAT). Dynamic Modulus were derived and Ultrasonic Pulse Velocity (UPV) tests were performed to validate RBT and SAT method results.

In general, Figure 1 shows that the curing time has a clear influence on the coating ability, abrasion resistance, and dynamic modulus especially at the initial stage of curing (within 28 days). Figure 1a,b show that the used fillers improved the bitumen coverage for both basalt and limestone aggregate compared with CE as a control filler, except SF which exhibited poor bitumen covering ability. It is worth noting that bitumen affinity to basalt aggregate is higher, especially at an early age, this finding is lined up with. When compared with CE, ET filler improved the bitumen coating ability after water erosion due to the early formed crystallin that increases the interlocking force between bitumen and aggregate surface, which improves adhesion between the mastic and the aggregate surface. In contrast, the bitumen coating ability of the CE specimen was considerably low. In the CE blended aggregate, the rigid hydration products improve the stiffness properties of the bitumen which, in turn, increases the stiffness of the mortar, as shown in Figure 1c, which improves the cohesion considerably but the adhesion slightly, and since the stripping resistance mostly depends on adhesion. Generally, all used fillers showed comparable abrasion resistance in 90 days of observation except SF. However, CE has slightly higher abrasion resistance on the first days of curing. Considering the effect of fillers on E, mortars with CE and ET exhibited the highest long-term and short-term performances, respectively. SF mortar performed the worst.

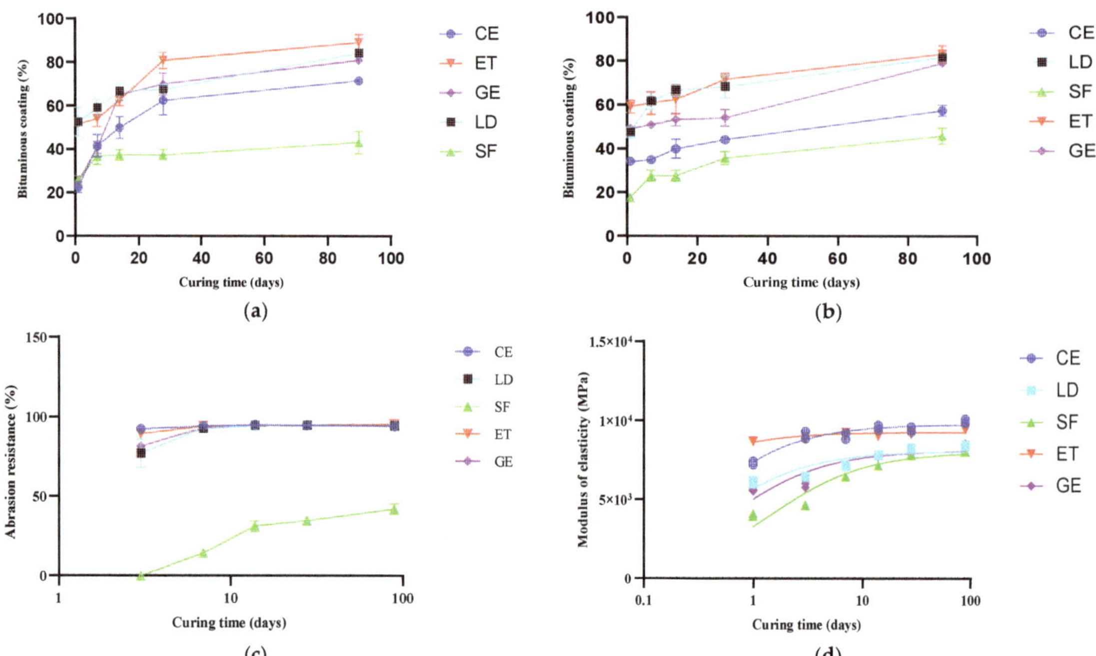

**Figure 1.** (**a**) Results of RBT test for the Basalt, (**b**) results of RBT test for the limestone, (**c**) results of SAT, and (**d**) results of UPV.

### 3. Conclusions and Recommendations

- Adding the active fillers provided a higher bitumen coverage and abrasion resistance than the SF, resulting in better affinity and moisture resistance, especially ET;
- The effect of filler on moisture sensitivity was found to be higher than the effect of aggregates;
- Adding ET filler provided higher E values at an early age, while the CE led to higher stiffening behavior in long term;
- LD and GO allowed for general lower stiffness and higher bitumen coverage and comparable abrasion resistance compared with CE;
- The result of the E test is generally correlated with abrasion resistance;
- Applying those methods and tests will provide a more comprehensive view for evaluating the moisture resistance of the CRM mixtures.

**Author Contributions:** Conceptualization, A.A.-M. and K.M.; methodology, A.A.-M.; investigation, A.A.-M.; writing—original draft preparation, A.A.-M.; writing—review and editing, A.A.-M.; visualization, A.A.-M.; supervision, K.M.; project administration, K.M.; funding acquisition, K.M. All authors have read and agreed to the published version of the manuscript.

**Funding:** This research was funded by the German Academic Exchange Service (DAAD).

**Institutional Review Board Statement:** Not applicable.

**Informed Consent Statement:** Not applicable.

**Data Availability Statement:** No new data were created or analyzed in this study. Data sharing is not applicable to this article.

**Conflicts of Interest:** The authors declare no conflict of interest.

## References


1. Al-Mohammedawi, A.; Mollenhauer, K. Characterization of mechanical properties and shrinkage behavior of Cold recycled material (CRM) stabilized with different active fillers. In Proceedings of the 7th International Conference on Road and Rail Infrastructure, Pula, Croatia, 11–13 May 2022.
2. Al-Mohammedawi, A.; Mollenhauer, K. A Synergic Study on The Fatigue-Fracture Behavior of Cold Recycling Materials Using Innovative Green Additives. In Proceedings of the 8th Transport Research Arena Conference, Lisbon, Portugal, 14–17 November 2022.

*Abstract*

# A Digital Twin for Monitoring the Construction of a Wind Farm [†]

Alejandra Ospina-Bohórquez [1,*], Jorge López-Rebollo [2], Pedro Muñoz-Sánchez [2] and Diego González-Aguilera [2]

1   TIDOP Research Group, University of Salamanca, Hornos Caleros, 50, 05003 Ávila, Spain
2   Department of Cartographic and Land Engineering, Higher Polytechnic School of Ávila, University of Salamanca, Hornos Caleros, 50, 05003 Ávila, Spain; jorge_lopez@usal.es (J.L.-R.); pedroms@usal.es (P.M.-S.); daguilera@usal.es (D.G.-A.)
*   Correspondence: ale.ospina15@usal.es
†   Presented at the 1st International Online Conference on Infrastructures, 7–9 June 2022; Available online: https://ioci2022.sciforum.net/.

**Keywords:** digital twin (DT); building information modelling (BIM); as-built model; CAD/CAE model

---

**Citation:** Ospina-Bohórquez, A.; López-Rebollo, J.; Muñoz-Sánchez, P.; González-Aguilera, D. A Digital Twin for Monitoring the Construction of a Wind Farm. *Eng. Proc.* **2022**, 17, 3. https://doi.org/10.3390/engproc2022017003

Academic Editor: Jesús Balado Frías

Published: 2 May 2022

**Publisher's Note:** MDPI stays neutral with regard to jurisdictional claims in published maps and institutional affiliations.

Digital twins (DTs) represent an emerging technology that can allow interaction between physical assets and their virtual replicas. These virtual replicas enclose the geometry derived from complex modelling procedures and the dynamism derived from artificial intelligence. Nowadays, DT applications are found in almost every engineering area, DTs serve different purposes, e.g., they test how new devices behave under diverse conditions or while being controlled, and monitor existing processes to help them improve.

The Building Information Modelling (BIM) methodology, for its part, has revolutionized and changed the construction engineering and architecture sector in recent times. BIM refers to a collaborative work methodology used for the conception and management of building and civil works projects that include a digital model that centralizes all the information (e.g., geometric, costs, maintenance, etc.). BIM models are theoretical and are derived from the design phase where this methodology is applied. Instead, the As-Built models refer to the representation of the actual work progress at each moment and reflect the reality and evolution of the construction site through time.

With the improvement in artificial intelligence, in terms of software capabilities for 3D modelling and simulation in construction environments (BIM models) related to Computer-Aided Design and Engineering (CAD/CAE) and Geographical Information System (GIS) technologies, DTs now have a place in urban projects, land management, and public infrastructure. However, until now, the use of DTs in this area has been limited as, in most cases, they are only used for high-quality 3D digital representation without connecting to other systems, dynamic analysis, or simulation.

This work proposes the creation of a DT for monitoring the construction of a wind farm. It draws a comparison between the BIM model (which contains the construction specifications) and the As-Built models that represent the actual construction at different times. It also helps to control deviations regarding civil works that may occur during construction. All the data obtained (position of the wind turbines, the platform of the footing, the trace of the road, the width of the roadway, the slope of the road, etc.) must be stored in order to be displayed in the most didactic way possible so that the user can clearly understand it. Then, the DT includes a connection to a database to obtain the necessary information for the 3D representation. The model comparison must be displayed according to what the user considers relevant in each case, e.g., delivering the BIM model and the As-Built model in a specific construction area.

The authors propose using the Unreal Engine to create an interface for user interaction that includes CAD/CAE models obtained from the BIM and As-Built models corresponding to different steps during the construction. Furthermore, the use of non-relational databases (MongoDB) is proposed since the data to be stored are semi-structured (not all areas of a model will have the same parameters), and the project needs are unpredictable since they can change alongside progression. The flexibility of non-relational databases can allow these variations to be efficiently captured without making significant changes to the database structure.

**Author Contributions:** Conceptualization and methodology, A.O.-B., J.L.-R., P.M.-S. and D.G.-A.; software, A.O.-B., J.L.-R. and P.M.-S.; validation, A.O.-B., J.L.-R., P.M.-S. and D.G.-A.; writing—original draft preparation A.O.-B.; writing—review and editing, A.O.-B., J.L.-R., P.M.-S. and D.G.-A. All authors have read and agreed to the published version of the manuscript.

**Funding:** This research received no external funding.

**Institutional Review Board Statement:** Not applicable.

**Informed Consent Statement:** Not applicable.

**Conflicts of Interest:** The authors declare no conflict of interest.

*engineering proceedings*

*Abstract*

# Development of Geopolymer Binders with Mixed Construction and Demolition Waste-Based Materials [†]

**Gurkan Yildirim** [1,2,*], **Ashraf Ashour** [1], **Emircan Ozcelikci** [2,3], **Muhammed Faruk Gunal** [2] **and Behlul Furkan Ozel** [4]

1    Department of Civil and Structural Engineering, University of Bradford, Bradford BD7 1DP, UK; a.f.ashour@bradford.ac.uk
2    Department of Civil Engineering, Hacettepe University, Ankara 06230, Turkey; eozcelikci@hacettepe.edu.tr (E.O.); farukgunal@hacettepe.edu.tr (M.F.G.)
3    Institute of Science, Hacettepe University, Ankara 06230, Turkey
4    Department of Civil Engineering, Yozgat Bozok University, Yozgat 66200, Turkey; b.furkan.ozel@bozok.edu.tr
*    Correspondence: g.yildirim@bradford.ac.uk
†    Presented at the 1st International Online Conference on Infrastructures, 7–9 June 2022; Available online: https://ioci2022.sciforum.net/.



**Citation:** Yildirim, G.; Ashour, A.; Ozcelikci, E.; Gunal, M.F.; Ozel, B.F. Development of Geopolymer Binders with Mixed Construction and Demolition Waste-Based Materials. *Eng. Proc.* **2022**, *17*, 4. https://doi.org/10.3390/engproc2022017004

Academic Editor: Frank Dehn

Published: 2 May 2022

**Publisher's Note:** MDPI stays neutral with regard to jurisdictional claims in published maps and institutional affiliations.



**Abstract:** As a consequence of the ever-increasing urban population and continuous development of industrialization and economies of the countries around the world, the construction and demolition industry has gained eye-catching popularity, although it is also considered one of the largest producers of solid wastes globally. In an effort to counteract the negative effects of the growing construction and demolition waste (CDW) issue, the current study focuses on the utilization of mixed CDW-based materials such as hollow brick (HB), red clay brick (RCB), roof tile (RT), glass (G) and concrete (C) in the production of geopolymer binders. These materials were acquired from demolished residential buildings in an urban transformation area and then subjected to an identical two-step crushing–milling procedure to reach sufficient fineness for geopolymerization. In the first stage of the study, these materials were used singly in the production of geopolymer binders to analyse the effects of material characteristics (e.g., fineness, chemical composition and crystalline nature) on the geopolymerization performance. Thereafter, these materials were used altogether in a quinary mixture to produce geopolymer binders with the purpose of better simulating the real-life conditions where CDWs are obtained altogether and are time-/energy-consuming to separate. In order to characterize the performance of different CDW-based materials, several mixture designs were made using sodium hydroxide (NaOH) as the alkali activator. After applying thermal curing to the geopolymer pastes, compressive strength tests were performed in addition to microstructural analyses. The results showed that compressive strength values of up to 55 MPa could successfully be achieved depending on the mixture proportions. While RT was found to be the most effective material in terms of the mechanical performance of CDW-based geopolymer binders, G and C exhibited poor performances due to relatively coarse particle size distribution and an inadequate chemical composition of $SiO_2$ and $Al_2O_3$, which is a necessity for effective geopolymerization. In-depth microstructural analyses identified that geopolymer pastes with higher compressive strengths had denser and more homogeneous microstructures. The main reaction products of the geopolymer binders were mostly sodium aluminosilicate hydrate (N-A-S-H) gels with zeolite-like structures, as well as some calcium aluminosilicate hydrate (C-A-S-H) gels that arose from the use of C with a high CaO content. Our results prove that CDW-based materials can successfully be used in the production of geopolymers, and can be regarded as promising alternatives to traditional systems based on Portland cement.

**Keywords:** construction and demolition waste (CDW); geopolymer; compressive strength; microstructure

**Author Contributions:** Conceptualization, G.Y. and A.A.; methodology, G.Y. and A.A.; software, G.Y.; validation, G.Y. and A.A.; formal analysis, G.Y. and A.A.; investigation, G.Y., E.O., M.F.G. and B.F.O.; resources, G.Y. and A.A.; data curation, G.Y., E.O., M.F.G. and B.F.O.; writing—original draft preparation, G.Y., E.O., M.F.G. and B.F.O.; writing—review and editing, G.Y. and A.A.; visualization, E.O.; supervision, A.A.; project administration, G.Y. and A.A.; funding acquisition, G.Y. and A.A. All authors have read and agreed to the published version of the manuscript.

**Funding:** This project has received funding from the European Union's Horizon 2020 research and innovation programme under the Marie Skłodowska-Curie grant agreement No [894100].

**Institutional Review Board Statement:** Not applicable.

**Informed Consent Statement:** Not applicable.

**Data Availability Statement:** The data presented in this study are available on request from the corresponding author. The data are not publicly available due to ongoing research and will be made available upon full publishment of the results.

**Conflicts of Interest:** The authors declare no conflict of interest.

*engineering proceedings*

*Abstract*

# Transform Physical Assets to 3D Digital Models [†]

**Linh Truong-Hong** 

Department of Geoscience and Remote Sensing, Delft University of Technology, 2628 CN Delft, The Netherlands; l.truong@tudelft.nl

† Presented at the 1st International Online Conference on Infrastructures, 7–9 June 2022; Available online: https://ioci2022.sciforum.net/.

**Keywords:** point cloud; contextual knowledge; surface extraction; digital twin; infrastructure inspection

**check for updates**

**Citation:** Truong-Hong, L. Transform Physical Assets to 3D Digital Models. *Eng. Proc.* **2022**, *17*, 5. https://doi.org/10.3390/engproc2022017005

Academic Editor: Joaquín Martínez-Sánchez

Published: 6 May 2022

**Publisher's Note:** MDPI stays neutral with regard to jurisdictional claims in published maps and institutional affiliations.



It is clearly a huge benefit for infrastructure monitoring, inspection, and management when a digital twin (DT) is developed to represent a real physical infrastructure. Three-dimensional (3D) geometric models of physical assets, particularly as-is 3D models, is the backbone of the DT, which are used to integrate real-time information of the physical assets and are fundamental components for modelling and simulation to predict responses of infrastructure. In the DT concept, the digital model requires to automatically update changes of physical infrastructure in an accurate and timely manner. Today, laser scanning sensors and cameras integrated into laser scanners, drones, and other surveillance equipment allow us to capture 3D topographic information of objects' surfaces in a 3D space with different level of detail and accuracy. As such, 3D point clouds representing to surface information of infrastructure derived from these surveying tools are to be a fundamental resource in creating 3D geometric models for DT. Automatically generating digital models from the 3D point clouds presents many challenges due to adverse quality and quantity of data points, massive data points, and highly complex geometries of 3D objects and scenes [1]. Moreover, in practice, existing workflows to achieve detailed, precise 3D geometric models of physical assets are mostly based on human work, implying high time consumption, cost, and possibility of human error. This paper proposes a framework using both spatial information of point clouds and contextual knowledge of objects to automatically extract point clouds of individual surfaces of objects of infrastructure (e.g., buildings and bridges). Contextual knowledge can include lower and upper bounds of dimensions of the objects, and a geometric relationship with adjoined objects. The main goal of the use of contextual knowledge is to support in estimating input parameters, to roughly extract point clouds of interest, and to filter unrealistic objects to be recognized. By integrating contextual knowledge into the framework, only a subset containing the point cloud of each object of interest needs be processed to extract the surfaces, the proposed framework can handle large bridge data sets. Once the point cloud of individual surfaces of each structural component are available, the 3D models of the structure can be created, or surface damage can be identified. Buildings and bridges are selected as case studies to demonstrate the proposed framework [2,3].

**Funding:** This study was funded by the generous support of the European Commission through H2020 MSCA-IF, "BridgeScan: Laser Scanning for Automatic Bridge Assessment", Grant 799149.

**Institutional Review Board Statement:** Not applicable.

**Informed Consent Statement:** Not applicable.

**Acknowledgments:** The authors also thank Dat Hop Company Limited, Ceotic., JSC and GeoInstinct Vietnam for their providing the laser scanning data.

**Conflicts of Interest:** The authors declare no conflict of interest. The funders had no role in the design of the study; in the collection, analyses, or interpretation of data; in the writing of the manuscript, or in the decision to publish the results.

## References

1. Truong-Hong, L.; Laefer, D.F. Quantitative evaluation strategies for urban 3D model generation from remote sensing data. *Comput. Graph.* **2015**, *49*, 82–91. [CrossRef]
2. Truong-Hong, L.; Lindenbergh, R. Automatically extracting surfaces of reinforced concrete bridges from terrestrial laser scanning point clouds. *Autom. Constr.* **2022**, *135*, 104127. [CrossRef]
3. Truong-Hong, L.; Lindenbergh, R. Extracting structural components of concrete buildings from laser scanning point clouds from construction sites. *Adv. Eng. Inform.* **2022**, *51*, 101490. [CrossRef]

*Abstract*

# Autogenous Healing in 10-Years Aged Cementitious Samples Containing Microfibers and Superabsorbent Polymers [†]

**Didier Snoeck** 

BATir Department, Université Libre de Bruxelles (ULB), 50 F.D. Roosevelt Ave., CP 194/02, 1050 Brussels, Belgium; didier.snoeck@ulb.be

† Presented at the 1st International Online Conference on Infrastructures, 7–9 June 2022; Available online: https://ioci2022.sciforum.net/.


**Keywords:** durability; sustainability; sustainable structural design; further hydration; calcium carbonate crystallization; self-healing; hydrogel; strain-hardening; age; decade

**Citation:** Snoeck, D. Autogenous Healing in 10-Years Aged Cementitious Samples Containing Microfibers and Superabsorbent Polymers. *Eng. Proc.* 2022, 17, 6. https://doi.org/10.3390/engproc2022017006

Academic Editor: Vyacheslav R. Falikman

Published: 2 May 2022

**Publisher's Note:** MDPI stays neutral with regard to jurisdictional claims in published maps and institutional affiliations.

## 1. Overview and Novelty

Due to the interest in increasing the durability and sustainability of concrete structures and construction techniques, a wide range of novel cementitious materials are being designed and investigated. One such recent material is a cementitious material containing superabsorbent polymers (SAPs) studied only from 1999 [1] onwards, mainly for its internal curing purposes with mitigation of autogenous shrinkage [2] and sealing characteristics [1]. Other positive influences are the change in rheology and the increase in freeze–thaw resistance, amongst others [3,4]. From 2010 onwards [5], a combination of addition of synthetic microfibers and SAPs was studied for their improved influence on autogenous healing in cementitious materials. It was found that optimal self-healing features were possible [6,7], as the crack widths were limited and water was available during dry periods. Some of those first samples now have an age of over 10 years.

As the autogenous-healing capacity is dependent on the age of the material, so will be the possible influence of added materials to promote this healing. The effects beyond one year [8] are not omnipresent in the literature. The effect of the age cannot be investigated as long as the actual specimens do not reach the required maturity. In a previous study, the age was studied up to 8 years' time [9]. In this study, specimens from the same batch were studied after a decade of maturing in different storage conditions.

## 2. Methodology and Results

Samples containing CEM I 52.5 N (1:1), fly ash (1:1), fine quartz sand (0.7:1), water (0.6:1), superplasticizer (0.01:1), PVA fibers (0.04:1) were used as reference. The SAP samples contained additionally SAPs (0.01:1) and extra water (0.09:1) on top. The SAP is a bulk-polymerized cross-linked potassium salt polyacrylate with a $d_{50}$ particle size of 477 µm and can swell up to 300 times its own weight in a liquid [7].

Samples were prepared and stored for 28 days at $20 \pm 2\ °C$ and a RH > 95%. Three storage conditions up to an age of 10 years were used. These were (1) $20 \pm 2°C$ and an RH > 95%, (2) a standard laboratory condition of $20 \pm 2\ °C$ and an RH of $60 \pm 5$%, and (3) exposed outdoor storage in a Belgium climate 5 km from the weather station in Melle.

The effect of autogenous healing was investigated by four-point-bending loading at the age of 10 years. First, specimens were loaded to 1% strain. Second, the samples were stored in specific healing conditions. These were the same as pre-conditioning with one additional (4) wet-dry cycling with 1 h submersion in water at $20 \pm 0.5\ °C$ and 23 h storage in standard laboratory conditions of $20 \pm 2\ °C$ and an RH of $60 \pm 5$%. After this curing period, the samples were reloaded up to 1% strain. The results are given in Table 1.

**Table 1.** First-cracking strength $\sigma_{fc}$ [MPa], average crack width $w$ [µm], number of cracks # [-], and healing ratio HR at high *RH+*, standard *RH-*, outdoor conditions *out* and wet/dry cycling *wd* [%].

| Sample | $\sigma_{fc}$ | $w$ | # | *HR RH+* | *HR RH-* | *HR out* | *HR wd* |
|---|---|---|---|---|---|---|---|
| REF | $6.5 \pm 0.5$ MPa | $10 \pm 7$ µm | 2–8 | $5 \pm 1\%$ | $0 \pm 1\%$ | $9 \pm 6\%$ | $15 \pm 9\%$ |
| SAP | $6.0 \pm 0.8$ MPa | $8 \pm 6$ µm | 4–12 | $16 \pm 6\%$ | $6 \pm 8\%$ | $27 \pm 8\%$ | $33 \pm 7\%$ |

Typical strengths and crack widths were obtained. Due to the stress initiator property of SAPs [10], the number of cracks increases. Due to the macropore formation, the strength is lowered. However, the healing ratios are always higher for SAP compared to REF samples. This is due to the water action by the SAPs during dry periods and the ability of SAPs to extract moisture from the ambient environment. This leads to better conditions for healing products to form as water is available [11]. The main visual appearance of the healing products was the whitish calcium carbonate crystallization.

### 3. Conclusions and Recommendations

The small crack widths after 10 years are still able to be partially healed. The main visual healing product is calcium carbonate. Further hydration was less likely as most of the binder already hardened during storage conditions. Generally, the samples containing SAPs show more prominent healing and they are still able to swell almost completely after a decade of storage in an alkaline cementitious environment. This makes them a sustainable option for the future as less maintenance and repair will be required.

**Funding:** This research received no external funding.

**Institutional Review Board Statement:** Not applicable.

**Informed Consent Statement:** Not applicable.

**Data Availability Statement:** The datasets generated during and/or analyzed during the current study are available from the corresponding author on reasonable request.

**Acknowledgments:** The author wants to thank the companies for providing the materials.

**Conflicts of Interest:** The author declares no conflict of interest.

## References

1. Tsuji, M.; Shitama, K.; Isobe, D. Basic studies on simplified curing technique, and prevention of initial cracking and leakage of water through cracks of concrete by applying superabsorbent polymers as new concrete admixture. *J. Soc. Mater. Sci.* **1999**, *48*, 1308–1315. [CrossRef]
2. Jensen, O.M.; Hansen, P.F. Water-entrained cement-based materials I. Principles and theoretical background. *Cem. Concr. Res.* **2001**, *31*, 647–654. [CrossRef]
3. Mechtcherine, V.; Wyrzykowski, M.; Schröfl, C.; Snoeck, D.; Lura, P.; De Belie, N.; Mignon, A.; Van Vlierberghe, S.; Klemm, A.J.; Almeida, F.C.R.; et al. Application of super absorbent polymers (SAP) in concrete construction—Update of RILEM state-of-the-art report. *Mater. Struct.* **2021**, *54*, 80. [CrossRef]
4. Schröfl, C.; Erk, K.A.; Siriwatwechakul, W.; Wyrzykowski, M.; Snoeck, D. Recent progress in superabsorbent polymers for concrete. *Cem. Concr. Res.* **2022**, *151*, 106648. [CrossRef]
5. Kim, J.S.; Schlangen, E. *Super Absorbent Polymers to Simulate Self Healing in ECC*; van Breugel, K., Ye, G., Yuan, Y., Eds.; RILEM Publications SARL: Delft, The Netherlands, 2010; pp. 849–858.
6. Snoeck, D. Superabsorbent polymers to seal and heal cracks in cementitious materials. *RILEM Tech. Lett.* **2018**, *3*, 32–38. [CrossRef]
7. Snoeck, D.; Van Tittelboom, K.; Steuperaert, S.; Dubruel, P.; De Belie, N. Self-healing cementitious materials by the combination of microfibres and superabsorbent polymers. *J. Intell. Mater. Syst. Struct.* **2014**, *25*, 13–24. [CrossRef]
8. Yıldırım, G.; Khiavi, A.H.; Yeşilmen, S.; Şahmaran, M. Self-healing performance of aged cementitious composites. *Cem. Concr. Compos.* **2018**, *87*, 172–186. [CrossRef]
9. Snoeck, D.; De Belie, N. Autogenous healing in strain-hardening cementitious materials with and without superabsorbent polymers: An 8-year study. *Front. Mater.* **2019**, *6*, 48. [CrossRef]

10.	Yao, Y.; Zhu, Y.; Yang, Y. Incorporation of SAP particles as controlling pre-existing flaws to improve the performance of ECC. *Constr. Build. Mater.* **2011**, *28*, 139–145. [CrossRef]
11.	Snoeck, D.; Pel, L.; De Belie, N. Autogenous Healing in Cementitious Materials with Superabsorbent Polymers Quantified by Means of NMR. *Sci. Rep.* **2020**, *10*, 642. [CrossRef]

MDPI

*Abstract*

# Graph-Based Version Control of BIM Models in an Event-Driven Collaboration Environment †

**Sebastian Esser** 

School of Engineering and Design, Technical University of Munich, 80333 Munich, Germany;
sebastian.esser@tum.de
† Presented at the 1st International Online Conference on Infrastructures, 7–9 June 2022; Available online:
https://ioci2022.sciforum.net/.

**Keywords:** asynchronous collaboration; object-based version control; BIM level 3; vendor-neutral
data exchange; common data environment



**Citation:** Esser, S. Graph-Based
Version Control of BIM Models in an
Event-Driven Collaboration
Environment. *Eng. Proc.* **2022**, *17*, 7.
https://doi.org/10.3390/
engproc2022017007

Academic Editor: Joaquín
Martínez-Sánchez

Published: 5 May 2022

**Publisher's Note:** MDPI stays neutral
with regard to jurisdictional claims in
published maps and institutional affiliations.



Interdisciplinary collaboration and communication are two essential aspects of Building Information Modeling (BIM). Current practice and international standards rely on
exchanging entire domain models, which are managed as separated files and coordinated
in a primarily manual fashion. The concept lacks version control, as the granularity of
change tracking remains on the level of complete monolithic files. Hence, high manual
effort is necessary to coordinate model modifications across the domains involved in
a project.

To overcome the limitations addressed, the keynote presents a novel approach that
enables modification tracking on object level instead of tracking monolithic model files. As
BIM models contain not only objects but also various dependencies forming a complex
network structure, formalisms of graph theory and graph transformation are applied to
identify and deploy model changes in a vendor- and schema-neutral fashion [1]. The
communication among project partners is ultimately implemented using event-driven
network architectures, which provide a flexible means to realize scalable asynchronous
collaboration [2]. Once an authoring party reaches a new shareable state of its discipline
model, an update event is raised and deployed through a central project hub. Each event
contains a set of transformation rules and additional information relevant to project management purposes. Applying the transformations to an outdated model copy, concurrency
among all existing replicas of a particular discipline model is obtained again. As a key
advantage, the updates are much smaller compared to repeatedly exchanging entire BIM
models. Furthermore, the approach provides a responsive and scalable system where
each design unit can subscribe to specific events like modifications of specific object types
or models of a particular discipline. Finally, the approach fits into existing standards of
model-based collaboration such as ISO 19650 or the concept of Information Containers for
linked Document Delivery (ICDD) defined in ISO 21597.

The application of the proposed collaboration environment is demonstrated using
BIM models implementing the Industry Foundation Classes (IFC) as their underlying
data model.

**Funding:** This research received no external funding.

**Institutional Review Board Statement:** Not applicable.

**Informed Consent Statement:** Not applicable.

**Conflicts of Interest:** The authors declare no conflict of interest.

## References

1. Esser, S.; Vilgertshofer, S.; Borrmann, A. Graph-based version control for asynchronous BIM level 3 collaboration. In Proceedings of the EG-ICE 2021 Workshop on Intelligent Computing in Engineering, Berlin, Germany, 30 June–2 July 2021; pp. 98–107. [CrossRef]
2. Esser, S.; Abualdenien, J.; Vilgertshofer, S.; Borrmann, A. Requirements for event-driven architectures in open BIM collaboration. In Proceedings of the 29th International Workshop on Intelligent Computing in Engineering, Aarhus, Denmark, 6–8 July 2022.

*Abstract*

# Influence of Precursor Materials on the Mechanical Behavior of Ambient-Cured One-Part Engineered Geopolymer Composites [†]

**Wee Teo** [1,*] , **Kazutaka Shirai** [2] **and Jee Hock Lim** [3]

1　School of Energy, Geoscience, Infrastructure and Society (EGIS), Heriot-Watt University Malaysia, Putrajaya 62200, Malaysia
2　Faculty of Engineering, Hokkaido University, Sapporo 060-8628, Japan; shirai.kazutaka@eng.hokudai.ac.jp
3　Department of Civil Engineering, Universiti Tunku Abdul Rahman, Bandar Sungai Long, Cheras, Kajang 43000, Malaysia; limjh@utar.edu.my
*　Correspondence: t.wee@hw.ac.uk; Tel.: +60-127560230
†　Presented at the 1st International Online Conference on Infrastructures, 7–9 June 2022; Available online: https://ioci2022.sciforum.net/.

**Keywords:** geopolymer; Engineered Geopolymer Composites; EGC; one-part; ambient-cured

**Citation:** Teo, W.; Shirai, K.; Lim, J.H. Influence of Precursor Materials on the Mechanical Behavior of Ambient-Cured One-Part Engineered Geopolymer Composites. *Eng. Proc.* **2022**, *17*, 8. https://doi.org/10.3390/engproc2022017008

Published: 2 May 2022

**Publisher's Note:** MDPI stays neutral with regard to jurisdictional claims in published maps and institutional affiliations.



Geopolymers are emerging low-carbon cement-free binders that offer a sustainable and environmentally friendly alternative to ordinary Portland cement (OPC). Despite their outstanding environmental friendliness, geopolymers still exhibit inherently brittle behavior similar to that of conventional cement-based concrete. Recently, there has been renewed interest in developing a new material combining geopolymers and Engineered Cementitious Composite (ECC) technologies, called Engineered Geopolymer Composites (EGCs). However, there are two major drawbacks associated with conventional geopolymer binders: firstly, it requires the handling of hostile, corrosive and viscous alkaline solutions; secondly, heat curing is necessary to improve the geopolymerisation process and mechanical properties. To overcome such limitations, a new class of geopolymer composites known as "one-part" or "just add water" geopolymers was developed for this purpose.

The concept of ECCs relies heavily on the micromechanics-based design principles, which provide a guide for tailoring of fiber, matrix and fiber–matrix interfaces to attain desired tensile ductility. Through careful tailoring, the fiber volume fraction usually remains moderate, at typically less than 2.5%. Polyvinyl alcohol (PVA) fiber is the most common type of fiber used in ECCs. To develop a cement-less EGC, a proper consideration of the geopolymer matrix design is essential. Research on EGCs is still relatively new. Preliminary feasibility studies carried out on slag-based EGCs [1] and fly-ash-based EGCs [2] have shown very promising results with a high tensile ductility over 4%. Studies on one-part EGCs conducted by Nematollahi et al. [3] and Alrefaei et al. [4] further assure more detailed investigations are needed for potential applications of this technology in future eco-friendly civil infrastructure.

This paper presents the results of a preliminary investigation on the influence of precursor materials on the mechanical properties of one-part EGCs. The aluminosilicate precursor materials used in this study consisted of a combination of fly ash (FA), ground granulated blast furnace slag (GGBS) and quartz powder (QP). Sodium metasilicate anhydrous was used as the solid alkali activator to synthesize the ambient-cured one-part geopolymer composites. In order to minimize the matrix fracture toughness, all mixtures were prepared without the addition of silica sand. All mixtures were designed with varying proportions of FA, GGBS and QP, amounts of alkali activators and water contents. Mechanical properties were determined by compression and direct tension tests. Fresh properties and microstructure analysis of each mixture were also studied and discussed.

The results indicate that the combination of GGBS with FA improves the reactivity of the mixture and compressive strength and enables a possible ambient curing condition.

Due to the spherical nature of FA's particle shape, the best ratio for the combination of FA and GGBS in terms of flowability was found to be 70:30. An increase in the number of solid alkali activators used, reduction in the water contents and addition of QP could beneficially increase the compressive strength as well as uniaxial tensile cracking strength, ultimate strength and strain capacity. This was clearly reflected in the microstructure of the geopolymer gel, which showed a more compact and denser morphology.

**Author Contributions:** Conceptualization, W.T.; methodology, W.T. and J.H.L.; validation, W.T., K.S. and J.H.L.; formal analysis, W.T.; investigation, W.T. and J.H.L.; resources, W.T.; writing—original draft preparation, W.T.; writing—review and editing, W.T., K.S. and J.H.L.; visualization, W.T.; supervision, W.T.; project administration, W.T.; funding acquisition, W.T. and K.S. All authors have read and agreed to the published version of the manuscript.

**Funding:** This research was funded by the Ministry of Education (MOE) Malaysia through the Fundamental Research Grant Scheme (FRGS), grant number: FRGS/1/2018/TK01/HWUM/02/3, and in part by the 2021 ASAHI GLASS Foundation Research Grant through with Hokkaido University, Japan.

**Acknowledgments:** The authors would like to thank former research assistant Woo Chee Zheng for his contribution to the laboratory work.

**Conflicts of Interest:** The authors declare no conflict of interest. The funders had no role in the design of the study; in the collection, analyses, or interpretation of data; in the writing of the manuscript; or in the decision to publish the results.

## References

1. Lee, B.Y.; Cho, C.-G.; Lim, H.-J.; Song, J.-K.; Yang, K.-H.; Li, V.C. Strain hardening fiber reinforced alkali-activated mortar—A feasibility study. *Constr. Build. Mater.* **2012**, *37*, 15–20. [CrossRef]
2. Ohno, M.; Li, V.C. A feasibility study of strain hardening fiber reinforced fly ash-based geopolymer composites. *Constr. Build. Mater.* **2014**, *54*, 163–168. [CrossRef]
3. Nematollahi, B.; Sanjayan, J.; Qiu, J.; Yang, E.-H. Micromechanics-based investigation of a sustainable ambient temperature cured one-part strain hardening geopolymer composite. *Constr. Build. Mater.* **2017**, *131*, 552–563. [CrossRef]
4. Alrefaei, Y.; Dai, J.-G. Tensile behaviour and microstructure of hybrid fiber ambient cured one-part engineered geopolymer composites. *Constr. Build. Mater.* **2018**, *184*, 419–431. [CrossRef]

*Abstract*

# Evaluation of Binder-Aggregate Adhesion in Hot-Recycled Asphalt Mixtures as a Function of the Production Temperature [†]

**Edoardo Bocci** [1,*], **Emiliano Prosperi** [2] and **Maurizio Bocci** [2]

[1]  Faculty of Engineering, Università eCampus, Via Isimbardi 10, 22060 Novedrate, Italy
[2]  Department of Construction, Civil Engineering and Architecture, Università Politecnica delle Marche, Via Brecce Bianche, 60131 Ancona, Italy; e.prosperi@pm.univpm.it (E.P.); m.bocci@univpm.it (M.B.)
*  Correspondence: edoardo.bocci@uniecampus.it; Tel.: +39-338-467-7744
†  Presented at the 1st International Online Conference on Infrastructures, 7–9 June 2022; Available online: https://ioci2022.sciforum.net/.

**Keywords:** asphalt concrete; reclaimed asphalt; hot recycling; adhesion; temperature

---

**Citation:** Bocci, E.; Prosperi, E.; Bocci, M. Evaluation of Binder-Aggregate Adhesion in Hot-Recycled Asphalt Mixtures as a Function of the Production Temperature. *Eng. Proc.* **2022**, *17*, 9. https://doi.org/10.3390/engproc2022017009

Academic Editor: Salvatore Mangiafico

Published: 2 May 2022

**Publisher's Note:** MDPI stays neutral with regard to jurisdictional claims in published maps and institutional affiliations.

When recycling reclaimed asphalt (RA) in new hot-mix asphalt (HMA), the temperature of the mix components (mainly virgin aggregate, RA, and virgin bitumen) can vary in a wide range [1]. Higher temperatures of virgin aggregate allow the mobilization of a higher amount of binder in RA. However, this implies a more severe short-term aging of the virgin bitumen and poorer properties of the aged-virgin bitumen blend, due to the lower virgin bitumen/RA bitumen ratio [2]. On the contrary, the adoption of lower temperatures has the opposite effect (lower mobilization of the RA binder, but higher performance of the bituminous blend). In addition, the reduction of material heating results in a lower bitumen viscosity, which may determine a lower compactability and lower adhesiveness. Previous studies showed that a reduction of 30 °C in the mixing temperature of HMA containing RA does not imply a significant increase of the air voids content but allows improving the material performance against cracking, fatigue, and rutting [3]. Moreover, the lower mixing temperature also preserves the effectiveness of the rejuvenating agent [4]. To have a deeper understanding of this phenomenon, the adhesive properties between binder and aggregate were investigated through the simulation of a hot recycled HMA production in the laboratory, adopting two mixing temperatures.

The objectives of the research were: (i) evaluating how the binder adhesive properties changes when varying the content of aged bitumen; (ii) assessing if the adhesion is higher on virgin aggregate or on RA particles, coated with aged bitumen; (iii) understanding how the blending temperature influence binder–aggregate adhesion.

To this aim, binder bond strength (BBS) tests were carried out using a self-aligning Pneumatic Adhesion Tensile Testing Instrument (PATTI), according to AASHTO TP-91. The experimental program provided two types of substrates, simulating virgin limestone aggregate and RA, three RA/virgin binder proportions (20/80, 35/65 and 50/50), two types of rejuvenator in the binder (coded with the letters A and B), two bitumen application temperatures (140 °C and 170 °C), and five repetitions. In particular, a 50/70 penetration bitumen was used as virgin binder. To reproduce the RA substrate, the virgin bitumen at 170 °C was spread on hot limestone plates with an average thickness of about 10 μm. Then, the plates were aged in an oven at 135 °C for 4 h and 85 °C for 120 h, according to AASHTO R30. The same 50/70 virgin binder was aged in the laboratory using RTFOT (163 °C for 85 min) and PAV (100 °C and 2.1 MPa for 20 h) devices to reproduce the RA bitumen. The aged and virgin binders were blended with ratios 20/80, 35/65 and 50/50 to investigate different mobilization rates of the RA bitumen. In this step, 9% by aged bitumen weight of rejuvenator (A or B) was added. Binder blending and pull-stub gluing were carried out at 140 °C or 170 °C. The BBS tests were performed at 25 °C.

Figure 1 shows the measured values of pull-off tensile strength (POTS).

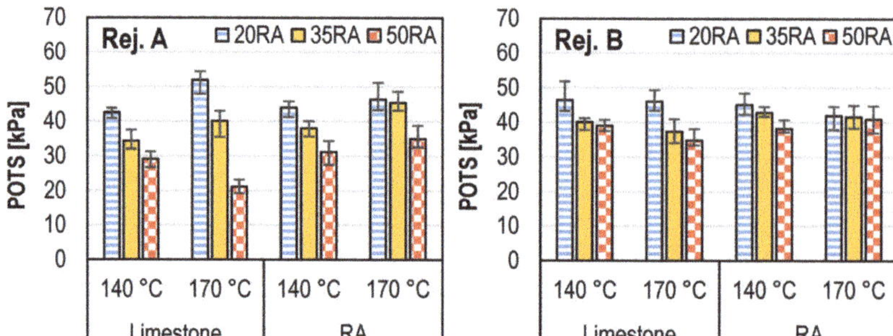

**Figure 1.** Pull-off tensile strength (POTS) values.

It can be immediately noted that the adhesive properties of the binder decreased when increasing the aged bitumen content from 20% to 50%. As the BBS tests provided quite dispersed data, a statistical analysis was carried out using a *t*-test. The $\alpha$ values obtained when comparing the POTS of the blends with 20% and 35% of RA bitumen and the blends with 35% and 50% of RA bitumen were, respectively, $1.2 \times 10^{-6}$ and $6.1 \times 10^{-7}$, confirming the decreasing trend of POTS with aged bitumen content. Moreover, the graphs in Figure 1 show that, for high RA bitumen contents (35% and 50%), the adhesion on the RA substrate was higher than on the limestone substrate ($\alpha = 0.004$). Between the two rejuvenators, the type B allowed obtaining higher POTS values for high RA bitumen contents (35% and 50%), as confirmed by $\alpha = 0.008$. Differently, the bitumen application temperature (140 °C or 170 °C) did not significantly influence the POTS ($\alpha = 0.50$). This indicates that the increase of adhesiveness that can be obtained at higher temperature was approximately balanced by the more severe aging underwent by the binder. However, as in site the lower mixing temperature implies the lower mobilization of the RA binder, thus a lower RA/virgin bitumen proportion, from the experimental results it can be stated that the reduction of the mix temperature is beneficial for the adhesion between the binder and both the virgin and the pre-coated RA aggregates.

The promising findings of the research encourages further studies on hot-recycling of RA at reduced mixing temperatures, also adopting warm mix asphalt solutions.

**Author Contributions:** Conceptualization, E.B. and M.B.; methodology, E.B.; formal analysis, E.P.; investigation, E.P.; data curation, E.B.; writing—original draft preparation, E.B.; writing—review and editing, E.P.; supervision, M.B. All authors have read and agreed to the published version of the manuscript.

**Funding:** This research received no external funding.

**Institutional Review Board Statement:** Not applicable.

**Informed Consent Statement:** Not applicable.

**Conflicts of Interest:** The authors declare no conflict of interest.

## References

1. Ma, X.; Leng, Z.; Wang, L.; Zhou, P. Effect of Reclaimed Asphalt Pavement Heating Temperature on the Compactability of Recycled Hot Mix Asphalt. *Materials* **2020**, *13*, 3621. [CrossRef] [PubMed]
2. Lo Presti, D.; Vasconcelos, K.; Orešković, M.; Menegusso Pires, G.; Bressi, S. On the Degree of Binder Activity of Reclaimed Asphalt and Degree of Blending with Recycling Agents. *Road Mater. Pavement* **2020**, *21*, 2071–2090. [CrossRef]
3. Marsac, P.; Bocci, E.; Cardone, F.; Cannone Falchetto, A.; Carbonneau, X.; Zaumanis, M.; Carter, A.; Rubio-Gámez, M.C.; Del Sol Sánchez, M.; Dave, E.V.; et al. International Evaluation of the Performance of Warm Mix Asphalts with High Reclaimed Asphalt Content. In Proceedings of the RILEM International Symposium on Bituminous Materials, Lyon, France, 14–16 December 2020; Volume 27, pp. 19–25. [CrossRef]
4. Prosperi, E.; Bocci, E.; Bocci, M. Evaluation of the Rejuvenating Effect of Different Additives on Bituminous Mixtures Including Hot-Recycled RA as a Function of the Production Temperature. *Road Mater. Pavement* **2021**, 1–20. [CrossRef]

*Abstract*

# Banana Fiber-Reinforced Geopolymer-Based Textile-Reinforced Mortar [†]

Vincent P. Pilien [1,*], Lessandro Estelito O. Garciano [1], Michael Angelo B. Promentilla [1], Ernesto J. Guades [2], Julius L. Leaño [3], Andres Winston C. Oreta [1] and Jason Maximino C. Ongpeng [1]

[1] College of Engineering, De La Salle University, Manila 0922, Philippines; lessandro.garciano@dlsu.edu.ph (L.E.O.G.); michael.promentilla@dlsu.edu.ph (M.A.B.P.); andres.oreta@dlsu.edu.ph (A.W.C.O.); jason.ongpeng@dlsu.edu.ph (J.M.C.O.)

[2] School of Engineering, University of Guam, Mangilao, GU 96923, USA; guadese@triton.uog.edu

[3] Research and Development Division, Philippine Textile Research Institute, Department of Science and Technology, Taguig City 1631, Philippines; jlleanojr@ptri.dost.gov.ph

[*] Correspondence: vincent_pilien@dlsu.edu.ph

[†] Presented at the 1st International Online Conference on Infrastructures, 7–9 June 2022; Available online: https://ioci2022.sciforum.net/.



**Abstract:** Textile-reinforced mortar (TRM) is an effective method for confining concrete elements to elevate the axial load resistance and upgrade the overall performance of concrete. TRM is a promising alternative to carbon-fiber-reinforced polymers (CFRP) which are commonly used to strengthen concrete and are known to be expensive since they require a huge amount of energy in processing these materials. Green technologies can be applied in this process, following the same TRM principles of confinement, replacing conventional cement or epoxy-based mortars and synthetic textiles towards sustainable concrete strengthening technology. This is through the utilization of a geopolymer mortar reinforced with short banana fibers (BF) and long BFs as textiles. Geopolymer mortar presented in this paper is composed of fly ash and silica fume as the binder, sand as the filler, sodium hydroxide (NaOH) and sodium silicate ($Na_2SiO_3$) as the activator and BFs as the reinforcement and textile. Geopolymerization generates significantly less carbon dioxide ($CO_2$) while BFs are known for having attractive mechanical properties, are cost effective and abundant in nature, and thus the use of this fiber will significantly minimize the huge waste produced from banana plantations after a one-time fruit harvest. The geotextile or geogrid used to wrap the concrete cylinder samples is made up of 2 mm-long BF yarns with weights ranging from 150 to 450 grams per square meter that varies with grid sizes from 10 mm, 15 mm to 25 mm for both orthogonal directions considering the lightweight characteristic of BFs. Twelve TRM designs were used to strengthen the concrete cylinders with three samples each. TRM design parameters vary in the thicknesses of the geopolymer mortar covering and the size of the geotextile grids. Eighteen of the geotextiles used were coated with a polymer to protect the fibers while the other eighteen geotextiles remained uncoated. A total of thirty-nine concrete cylinders with 150 mm base diameter and 300 mm height cured within 28 days were prepared, for which 36 cylinders were confined with green TRM with different parameters while three of the plain concrete cylinders served as the control specimens. This is to maximize the investigation on the potential of green TRM in confining concrete and to determine the variations in compressive strengths and mode of failures of confined and unconfined concrete specimens. Results highlighted notable enhancement in the mechanical properties of the modified plain concrete after 28 days of TRM curing using a universal testing machine (UTM). Likewise, a confinement theory of the optimum TRM design was modeled mathematically to evaluate the effects of concrete confinement and overall load carrying capacity enhancement gained from additional strength transferred by the TRM to the concrete element.

**Keywords:** green TRM; concrete confinement; natural fiber; modified concrete; compressive strength

**Author Contributions:** Conceptualization, V.P.P. and J.M.C.O.; methodology, V.P.P. and J.M.C.O.; software, V.P.P.; validation, V.P.P. and J.M.C.O.; formal analysis, V.P.P. and J.M.C.O.; investigation, V.P.P. and J.M.C.O.; resources, V.P.P., L.E.O.G., M.A.B.P., E.J.G., J.L.L., A.W.C.O. and J.M.C.O.; data curation, V.P.P. and J.M.C.O.; writing—original draft preparation, V.P.P.; writing—review and editing, V.P.P. and J.M.C.O.; visualization, V.P.P. and J.M.C.O.; supervision, V.P.P., L.E.O.G., M.A.B.P., E.J.G., J.L.L., A.W.C.O. and J.M.C.O.; project administration, V.P.P., L.E.O.G., M.A.B.P., E.J.G., J.L.L., A.W.C.O. and J.M.C.O.; funding acquisition, V.P.P. All authors have read and agreed to the published version of the manuscript.

**Funding:** This research received no external funding.

**Institutional Review Board Statement:** Not applicable.

**Informed Consent Statement:** Not applicable.

**Data Availability Statement:** Not applicable.

**Conflicts of Interest:** The authors declare no conflict of interest.

*Abstract*

# UAVs for Disaster Response: Rapid Damage Assessment and Monitoring of Bridge Recovery after a Major Flood [†]

**Marianna Loli [1,*], John Manousakis [2], Stergios A. Mitoulis [1] and Dimitrios Zekkos [3]**


[1] Department of Civil and Environmental Engineering, University of Surrey, Guildford GU2 7XH, UK; s.mitoulis@surrey.ac.uk
[2] Elxis Group, 11528 Athens, Greece; jmanousakis@elxisgroup.com
[3] Department of Civil and Environmental Engineering, University of California, Berkeley, CA 94720, USA; zekkos@berkeley.edu
* Correspondence: m.loli@surrey.ac.uk
† Presented at the 1st International Online Conference on Infrastructures, 7–9 June 2022; Available online: https://ioci2022.sciforum.net/.


**Keywords:** infrastructure; diagnosis; digital twins; resilience; natural hazards; flooding




**Citation:** Loli, M.; Manousakis, J.; Mitoulis, S.A.; Zekkos, D. UAVs for Disaster Response: Rapid Damage Assessment and Monitoring of Bridge Recovery after a Major Flood. *Eng. Proc.* **2022**, *17*, 11. https://doi.org/10.3390/engproc2022017011

Published: 2 May 2022

**Publisher's Note:** MDPI stays neutral with regard to jurisdictional claims in published maps and institutional affiliations.





While the planet is experiencing the roughest ecological disruption in our history, it is of utmost importance to try to mitigate the impact of intensifying natural disasters. Bridges are the priority for enabling climate resilience in transport infrastructure. They are inarguably the most valuable assets of transportation networks. Capital investment in bridge construction and maintenance in Europe is enormous, representing 30% of the total cost of transport networks. Nonetheless, bridges are too vulnerable. They are disproportionately exposed to natural hazards, especially floods, while becoming increasingly deficient due to ageing and urbanization trends.

Emerging Technologies are enablers of bridge resilience. Advocating for the use of UAVs in disaster response, this study provides solid and well-documented case studies discussing lessons learnt from the systematic analysis of field evidence after a recent (September 2020) Mediterranean Hurricane that struck central Greece. The use of UAVs proved essential for the rapid site reconnaissance and mapping of complex and severely damaged structures, including sinking piers, and collapsed abutments, with increased safety. UAVs effectively bypassed access blockages, resulting from failures in the road network, while, most importantly, allowing the execution of works with the minimum of human exposure to health risks during the peak of the COVID-19 pandemic.

The produced 3D models are powerful visualization tools that were found to fully compensate for the inability to physically visit the site, inspect and make decisions for severely damaged bridges. The value of this capability is also acknowledged by the researchers and engineers who performed the virtual inspections and who could not be present because of COVID-induced travel restrictions.

These models significantly facilitated the identification and analysis of the various bridge failure mechanisms, providing a uniquely comprehensive database of bridge response patterns under extreme flow velocities [1]. Furthermore, they proved useful as benchmarks for comparisons and informed decisions concerning the progress of restoration activities [2]. As such, they enabled accurate monitoring of bridge recovery and, incidentally, rapid assessment of the impact of an earthquake sequence that shook the region shortly after the flood, while mitigation works were underway. Thanks to this coincidence, we were given the opportunity to document a unique case study of the long-term performance of bridges in a multi-hazard environment, which should be of interest to all engineers and researchers in the field of civil infrastructure.

**Author Contributions:** M.L.: Conceptualization, methodology, validation, formal analysis, investigation, resources, data curation, writing—original draft preparation, writing—review and editing, visualization, project administration, funding acquisition; J.M.: Methodology, validation, formal analysis, investigation, resources, data curation, writing—review and editing; S.A.M.: Conceptualization, validation, resources, writing—review and editing, visualization, supervision, project administration, funding acquisition; D.Z.: Methodology, validation, resources, writing—review and editing, supervision. All authors have read and agreed to the published version of the manuscript.

**Funding:** This research was funded by the EUROPEAN UNION H2020-Marie Skłodowska-Curie Research Grants Scheme, grant number 895432.

**Institutional Review Board Statement:** Not applicable.

**Informed Consent Statement:** Not applicable.

**Conflicts of Interest:** The authors declare no conflict of interest.

## References

1. Loli, M.; Mitoulis, S.A.; Tsatsis, A.; Manousakis, J.; Zekkos, D. Flood characterization based on forensic analysis of bridge collapse using UAV reconnaissance and CFD simulations. *Sci. Total Environ.* **2022**, *822*, 153661. [CrossRef] [PubMed]
2. Mitoulis, S.A.; Argyroudis, S.A.; Loli, M.; Imam, B. Restoration models for quantifying flood resilience of bridges. *Eng. Struct.* **2021**, *238*, 112180. [CrossRef]

*engineering proceedings*

*Abstract*

# Defined-Performance Concretes Using Nanomaterials and Nanotechnologies [†]

Vyacheslav Falikman 

National Scientific Research Center Construction, Scientific Research Institute for Concrete and Reinforced Concrete after A.A. Gvozdev, 109428 Moscow, Russia; vfalikman@yahoo.com; Tel.: +7-985-764-8297
† Presented at the 1st International Online Conference on Infrastructures, 7–9 June 2022; Available online: https://ioci2022.sciforum.net/.

**Keywords:** defined-performance concrete; nanomaterials; nanotechnologies; sixth technological wave; smart properties; self-regulation



**Citation:** Falikman, V. Defined-Performance Concretes Using Nanomaterials and Nanotechnologies. *Eng. Proc.* **2022**, *17*, 12. https://doi.org/10.3390/engproc2022017012

Academic Editor: Vilma Ducman

Published: 2 May 2022

**Publisher's Note:** MDPI stays neutral with regard to jurisdictional claims in published maps and institutional affiliations.



The industry of building materials and construction, despite its obviously conservative character, quite often has to face the so-called "industrial revolution of the XXI century". New trends and new methods of experimentation and research are becoming the foundation for perspectives on the creation of high-tech products and processes characterized by a guaranteed reliability index, developing the principles for manufacturing up-to-date "supermaterials", and are marking the start of the sixth technological wave.

A special place among high-tech products is occupied by defined-performance concrete. An impressive breakthrough in construction technologies in the 21st century was achieved due to the properties of modern concrete, which have recently seemed unattainable. These include extremely low values of the water/cement ratio and air content of the concrete mixture, with long-lasting flowability, cohesion and uniformity; the ability for fresh concrete to easily and completely fill in a formwork of any configuration, with dense reinforcement, and without the use of energy for horizontal or vertical mix pouring; the ability for concrete to achieve a given strength, with the development of adjustable strength depending on the climatic factors; and a dense concrete structure at the nano-, micro- and macrolevel to ensure high strength, resistance and durability.

The interdisciplinary nature of concrete science contributes to large volumes of fundamental laws and the provisions of physical and colloid chemistry, chemistry of high-molecular-mass compounds, modeling methods, computer science, etc., being involved in their methodologies. Expanding the boundaries of understanding of its essence is an urgent task in modern concrete science.

All these concepts reflect the formation of a new technological pattern in concrete science and the concrete industry, which means a transition away from the established approaches and stereotypes.

The presence of nanomaterials and nanotechnologies in the construction segment is becoming more prominent. Today, in the total global market of nano-products, the construction industry "consumes" up to 3% of its volume and value in terms of the total market of nanomaterials, and in some segments, such as nanocomposites, up to 11%. The detailed analysis and long-term forecast for the development of research and the application of nanomaterials and nanotechnologies in construction shows that cement and concrete cover over 40% of the nanotechnology products in construction materials (the market value is about USD 5.6 billion), with a predicted annual growth of more than 10%.

In the transition from macro- to nano-range size, significant changes were noticed in electron conductivity, optical absorption, chemical reaction activity and mechanical properties, as well as in surface energy values and surface morphology of the composites.

The development of appropriate methods for determining properties and reaction control in nanostructures can lead to the creation of new materials, technologies and devices.

Recent advances in nano-chemistry and the development of new methods for the synthesis of nanoparticles are now expected to offer a new range of possibilities for the improvement of concrete performance. The incorporation of nanoparticles into conventional construction materials can provide the materials with advanced or smart properties that are of specific interest for high-rise, long-span or intelligent infrastructure systems.

Self-regulating concrete (SRC) is one of the most in-demand subjects in the modern concrete science. The choice of components and the design of SRC compositions are based on a prognostic assessment of the direction of spontaneous processes, to ensure high functionality at any technological or operational stage. The concept of "self-regulation" should be interpreted as the technologically predicted course of spontaneous processes in order to achieve the maximum possible functionality of the interacting components and concrete mixes, which meets the concept of defined-performance concrete (DPC).

Today, the successful implementation of a number of self-regulating concretes with defined performance is well known. Concretes that are self-compacting (self-consolidating), self-cleaning, self-healing, self-stressing and self-expanding, and self-sensing, and other much stronger, more rigid and durable structurally advanced cement materials stand out among them.

Examples of successful applications of $SiO_2$, $TiO_2$, $Fe_2O_3$, $Al_2O_3$, $CaCO_3$ nanoparticles; nanosized spinel $MgAl_2O_3$; nanoferrit $ZnFe_2O_4$; and nanoclays in concrete are given. The most promising contemporary developments include the synthesis and application of new forms of carbon, viz fullerene ($C_{60}$, $C_{70}$, $C_{540}$), graphene oxide (GO) and new types of carbon nanotubes.

For structural concrete, the most significant example of a wide industrial nanotechnology application is steel and FRC reinforcement with modified nanostructures. These bars have a much longer service life in a corrosive environment, which reduces the construction cost. Among the products, produced from the late 1990s on the basis of nanotechnologies, the most important are different coatings that increase the structural service life and give unique properties to structures.

Humankind is going through the changes in civilization's technical paradigm. Under the conditions of the planet's population growth and the inevitable emergence of raw material and power shortages in construction, quite a rapid displacement of traditional materials and technologies, through energy-saving and material-efficient solutions, must be a determining factor. Nano-binders and nano-engineered cement-based materials with nano-sized cementitious components, or other nano-sized particles, may be the next ground-breaking development.

In the near future, the manifestation of the general principles of nanotechnology for concrete and reinforced-concrete development should be expected in the production of high-quality ultra- and nanodispersed powders with stable chemical, phase and granulometric composition, in the development of new types of reinforcing elements (filamentary crystals, fibers, microspheres and dispersed particles); in the creation of new, defect-free, extremely strong reactive powder concretes, thermo-resistant composition materials with different electric conductivity levels, and nanosystems for health hazards and nuclear power stations; and in the development of the scientific foundations for designing specialized technology equipment with automated systems for cement-composite quality control.

**Funding:** This research received no external funding.

**Institutional Review Board Statement:** Not applicable.

**Informed Consent Statement:** Not applicable.

**Conflicts of Interest:** The author declares no conflict of interest.

*engineering proceedings*

*Abstract*

# A Framework for Intelligent Decision Making in Networks of Heterogeneous Systems (UAV and Ground Robots) for Civil Applications [†]

**Muhammad Imran** * and **Francesco Liberati**

Department of Computer, Control, and Management Engineering (DIAG), University of Rome, "La Sapienza", Via Ariosto, 25, 00185 Rome, Italy; liberati@diag.uniroma1.it
* Correspondence: imran@diag.uniroma1.it; Tel.: +39-3496867414
† Presented at the 1st International Online Conference on Infrastructures, 7–9 June 2022; Available online: https://ioci2022.sciforum.net/.



**Citation:** Imran, M.; Liberati, F. A Framework for Intelligent Decision Making in Networks of Heterogeneous Systems (UAV and Ground Robots) for Civil Applications. *Eng. Proc.* **2022**, *17*, 13. https://doi.org/10.3390/engproc2022017013

Academic Editor: Hosin Lee

Published: 2 May 2022

**Publisher's Note:** MDPI stays neutral with regard to jurisdictional claims in published maps and institutional affiliations.

**Abstract:** Cyber–physical systems (CPSs) are connected embedded devices with computing power, networking ability, control, and decision capability. The networks connecting these devices are different from the Internet because they can sense their environment, share information, make decisions, and act based on local and global information. These capabilities enable CPSs to improve processes in the transportation, agriculture, healthcare, and mining industries, and surveillance. Remarkable achievements in the development of cost effective, reliable, smaller, networked, and more powerful systems have allowed us to build new control and communication mechanisms, as well as cooperative and coordinated motion planning algorithms to enable these devices to assist humans to cope with real-time problems [1]. In this paper, we propose a learning-based distributed framework for intelligent decision making in networks of heterogeneous systems, to optimally plan their activities in highly dynamic environments. We utilized the multi-agent deep reinforcement learning (MADRL) technique to develop control and coordination strategies for teams of UAVs and group ground-moving robots. The developed framework enabled the team of unmanned aerial vehicles (UAVs) to observe the defined region above the ground correctly and efficiently, and to share information with ground robots, to perform robust actions. Our main objective was to maximize the utilization of the strong abilities of each CPS device. UAVs can observe the environment from above and rapidly gather reliable information to share with rescue robots working on the ground, but they cannot perform rescue tasks on the ground; in contrast, rescue robots cannot gather reliable information due to the lack of visual limitation. In this framework, we trained several DQN agents to learn optimal control policies for a team of cooperative heterogeneous robots in a centralized fashion, then perform actions in a decentralized way. These learned polices were further transferred in real time to the robots and evaluated against the real-time deployment of robots to perform tasks in the environment.

**Keywords:** cyber physical system; unmanned aerial vehicle; multi-agent reinforcement learning

---

**Author Contributions:** The two authors contributed to the research equally. All authors have read and agreed to the published version of the manuscript.

**Funding:** This research received no external funding.

**Institutional Review Board Statement:** Not applicable.

**Informed Consent Statement:** Not applicable.

**Data Availability Statement:** Not applicable.

---

**Conflicts of Interest:** The authors declare no conflict of interest.

## Reference

1.  Reisslein, M.; Rinner, B.; Roy-Chowdhury, A. Smart camera networks [guest editors' introduction]. *Computer* **2014**, *47*, 23–25.

*Abstract*

# Resource Efficiency to Achieve a Circular Economy in the Asphalt Road Construction Sector †

**Daniel Grossegger** 

School of Civil and Transportation Engineering, South China University of Technology, Guangzhou 510641, China; daniel1985@scut.edu.cn; Tel.: +86-18664804541

† Presented at the 1st International Online Conference on Infrastructures, 7–9 June 2022; Available online: https://ioci2022.sciforum.net/.

**Keywords:** resource efficiency; circular economy; asphalt roads; material flow analysis

**Citation:** Grossegger, D. Resource Efficiency to Achieve a Circular Economy in the Asphalt Road Construction Sector. *Eng. Proc.* **2022**, *17*, 14. https://doi.org/10.3390/engproc2022017014

Academic Editor: Amaryllis Audenaert

Published: 2 May 2022

**Publisher's Note:** MDPI stays neutral with regard to jurisdictional claims in published maps and institutional affiliations.

The construction and maintenance of the built environment consume a large quantity of resources and energy and contribute to the emission of a significant amount of greenhouse gases. Hence, improving resource efficiency and resource cycles is crucial to reducing environmental, economic and social impacts. The construction of roads mainly consumes mineral aggregates and binders, has the advantage of high recycling rates, and can utilise cascading materials. However, infinite recycling is impossible and recycled road construction material often cascades due to quality, quantity and economic issues. In addition, the extending and ageing road network faces increased traffic and climate changes, which might increase the probability of failure, inducing an increased maintenance effort. Maintenance has a minor contribution to resource consumption in developing countries, but it can have a major contribution in developed countries in the range of 50% to 75%. The increasing production of asphalt for surface wearing courses in Austria indicates the increase in materials used in maintenance. The production of surface course asphalt was about 25% to 35% before 2016, roughly reflecting the 3 cm asphalt surface layer of the total asphalt layer thickness of 15 cm to 20 cm used in municipalities' roads. The increase to 55% to 60% from 2017 to 2019 indicates the increasing maintenance work performed on the Austrian asphalt network. The reconstruction of roads and maintaining the road network accounted for about 65% in one Austrian municipality, reflecting the efforts of the municipality's administration to improve traffic concepts (increasing roadway width, adding cycle lanes and paths, reducing traffic speed and improving townscape), as well as addressing the structural problems and long-term solutions of degraded road surfaces. Since the reconstruction process is similar to initial construction, it consumes an identical amount of resources for asphalt layers. Hence, local factors such as traffic development, economic viability, and road lifespan are important to determine long-term resource efficiency. About 25% of the reclaimed asphalt of Austrian asphalt production in 2018 and 2020 corresponded to increased surface course asphalt production. This shows that system improvements are required to record waste generation, treatment and utilisation. 70% of the processed reclaimed asphalt is officially used to produce new asphalt. However, the cascading material flow of reclaimed asphalt pavements (used in unbound layers, gravel roads, road shoulders and backfilling) depends on local factors such as the short transportation distances of primary materials, low binder prices and administrative recycling commitments. A deeper understanding of the material flows related to asphalt roads, including primary and secondary material resources and resource consumers, and the economic interaction between the industries related to these flows, is necessary to establish sustainable asphalt roads without causing unwanted shifts in material flows and sustaining resource depletion.

**Funding:** This research was funded by the National Natural Science Foundation of China under the grant number 52150410430.

**Institutional Review Board Statement:** Not applicable.

**Informed Consent Statement:** Not applicable.

**Data Availability Statement:** The data presented in this study are available on request from the corresponding author.

**Conflicts of Interest:** The author declares no conflict of interest.

*Abstract*

# Drone-Image Based Fast Crack Analysis Algorithm Using Machine Learning for Highway Pavements [†]

**Byungkyu Moon and Hosin (David) Lee ***

Civil and Environmental Engineering, University of Iowa, Iowa City, IA 52242, USA; byungkyu-moon@uiowa.edu
* Correspondence: hosin-lee@uiowa.edu
† Presented at the 1st International Online Conference on Infrastructures, 7–9 June 2022; Available online: https://ioci2022.sciforum.net/.

**Keywords:** fast crack analysis algorithm; machine learning; highway pavements; digital image analysis; drone images

**Citation:** Moon, B.; Lee, H.(. Drone-Image Based Fast Crack Analysis Algorithm Using Machine Learning for Highway Pavements. *Eng. Proc.* **2022**, *17*, 15. https://doi.org/10.3390/engproc2022017015

**Academic Editor:** Joaquín Martínez-Sánchez

Published: 11 May 2022

**Publisher's Note:** MDPI stays neutral with regard to jurisdictional claims in published maps and institutional affiliations.

## 1. Introduction

Transportation agencies automatically collect and analyze pavement cracking data using agency-owned equipment and software or contracted services. These pavement cracking data are then used to determine the most appropriate maintenance and rehabilitation strategies to provide safe and reliable roadways [1]. However, this often requires high-cost equipment or services [2].

A digital image processing algorithm was developed to compute a unified crack index and crack type index [3,4]. A robust position invariant neural network was developed for digital pavement crack analysis [5]. The accuracy of automated pavement surface image analysis system has been evaluated against the ground-truth cracking data [6]. An image-based data collection procedure was then evaluated against the AASHTO provisional standard for cracking on asphalt-surfaced pavements [7].

Currently, ten state DOT's are using drones for bridge inspection and six state DOT's for pavement inspection [8]. Recently, there have been increased interests on automatically analyze drone images from integrators/service providers and end-users [9]. This paper presents a low-cost pavement distress data collection using a drone and subsequent drone image analysis using pavement crack analysis software.

This paper discusses state-of-the-art drone imaging technologies and advanced image analysis algorithms adopting advanced machine learning software tools. Drones were used to capture pavement surface images which were then analyzed using the crack image analysis software. This paper is timely given the increased new development in drone imaging technologies.

## 2. Methodology

Drone images were collected and a machine learning algorithm was developed for road segmentation and crack detection.

(1)    Data Set Preparation

Drone images of pavements were collected using a drone, which were then used for training for developing a machine learning algorithm. A second set of drone images were collected for validation of the developed machine learning algorithm.

(2)    Pavement Extraction from Drone Images

Drone images cover wide range of earth surface, and the first task is to extract pixels, which belong to pavements. To extract pavement pixels from drone images, a semantic segmentation method was used to develop a convolutional network architecture designed to accomplish the this first task.

(3)    Crack Detection

For a given crack image, a proposed machine learning algorithm was developed to yield a crack detection scheme, wherein the crack regions have a higher probability and non-crack regions have a lower probability. Figure 1 shows an example drone image acquisition and analysis result.

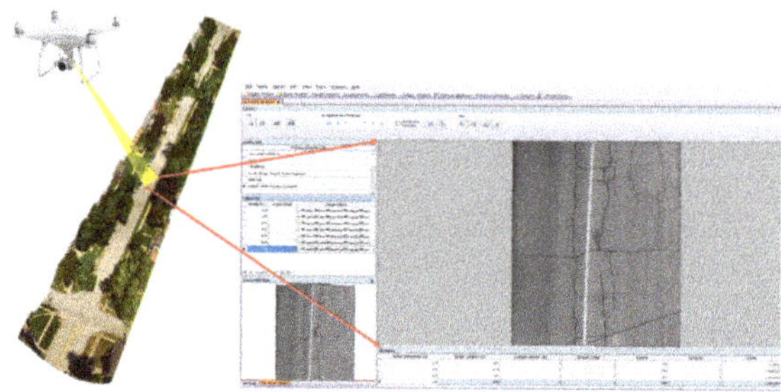

(**a**) Importing a drone image of pavement surface

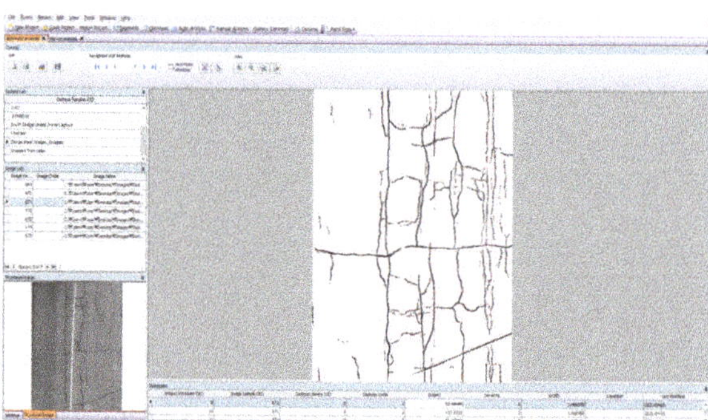

(**b**) Automatically analyzing a drone image

**Figure 1.** Importing and Analyzing a Drone image.

## 3. Summary and Conclusions

An increasing number of public agencies and companies are using drones for pavement inspection. Images can be automatically captured by a drone and stored in a point cloud for 3-D modeling. A DJI drone was used to capture pavement surface images in a high resolution at a low cost. Software was developed to analyze drone images and analysis results can be integrated with GIS software. In the future, an LiDAR camera can be mounted on a drone to measure a depth of cracks.

**Author Contributions:** Conceptualization, H.L.; methodology, B.M.; software, B.M.; validation, H.L.; formal analysis, H.L.; investigation, B.M.; resources, H.L.; data curation, B.M.; writing—original draft preparation, B.M.; writing—review and editing, H.L.; visualization, B.M.; supervision, H.L. All authors have read and agreed to the published version of the manuscript.

**Funding:** This research received no external funding.

**Institutional Review Board Statement:** Not applicable.

**Informed Consent Statement:** Not applicable.

**Data Availability Statement:** Not applicable.

**Conflicts of Interest:** The authors declare no conflict of interest.

## References

1. Wilde, W.J.; Thompson, L.; Wood, T.J. *Cost-Effective Pavement Preservation Solutions for the Real World*; (No. MN/RC 2014-33); Department of Transportation: Washington, DC, USA., 2014.
2. Albitres CM, C.; Smith, R.E.; Pendleton, O.J. Comparison of automated pavement distress data collection procedures for local agencies in San Francisco Bay Area, California. *Transp. Res. Rec.* **2007**, *1990*, 119–126. [CrossRef]
3. Jitprasithsiri, S.; Lee, H.; Sorcic, R.G. Development of Digital Image Processing Algorithm to Compute a Unified Crack Index for Salt Lake City. *Transp. Res. Rec.* **1996**, *1526*, 142–148. [CrossRef]
4. Lee, B.J.; Lee, H. A Robust Position Invariant Neural Network for Digital Pavement Crack Analysis. In *Computer-Aided Civil and Infrastructure Engineering*; Blackwell Publishing: Hoboken, NJ, USA, 2004; Volume 19, pp. 105–118.
5. Lee, H.; Kim, J. Development of crack type index. *Transp. Res. Rec.* **2005**, *1940*, 99–109. [CrossRef]
6. Lee, H.; Kim, J. Analysis of Errors in Ground Truth Indicators for Evaluating the Accuracy of Automated Pavement Surface Image Collection and Analysis System for Asset Management. *J. ASTM Int.* **2006**, *3*, 1–15. [CrossRef]
7. Raman, M.; Hossain, M.; Miller, R.; Cumberledge, G.; Lee, H.; Kang, K. Assessment of Image-Based data Collection and the AASHTO Provisional Standard for Cracking on Asphalt-Surfaced Pavements. *Transp. Res. Rec.* **2004**, *1889*, 116–125. [CrossRef]
8. Fischer, S.; Lu, J.; Van Fossen, K.; Lawless, E. *Global Benchmarking Study on Unmanned Aerial Systems for Surface Transportation: Domestic Desk Review*; (No. FHWA-HIF-20-091); Federal Highway Administration: Washington, DC, USA, 2020.
9. Mogawer, W.S.; Xie, Y.; Austerman, A.J.; Dill, C.; Jiang, L.; Gittings, J.E. *The Application of Unmanned Aerial Systems In Surface Transportation-Volume II-B: Assessment of Roadway Pavement Condition with UAS*; (No. 19-010); University of Massachusetts: Lowell, MA, USA, 2019.

*Abstract*

# The Monitoring Guidelines of the Lombardia Region in Italy [†]

**Maria Pina Limongelli** *[iD], **Carmelo Gentile** [iD], **Fabio Biondini** [iD], **Marco di Prisco, Francesco Ballio and Marco Belloli**

Department of Architecture, Built Environment and Construction Engineering, Politecnico di Milano, 20133 Milano, Italy; carmelo.gentile@polimi.it (C.G.); fabio.biondini@polimi.it (F.B.); marco.diprisco@polimi.it (M.d.P.); francesco.ballio@polimi.it (F.B.); marco.belloli@polimi.it (M.B.)

* Correspondence: mariagiuseppina.limongelli@polimi.it

† Presented at the 1st International Online Conference on Infrastructures, 7–9 June 2022; Available online: https://ioci2022.sciforum.net/.

**Keywords:** standardization; guidelines; structural health monitoring; dynamic and static monitoring; hydraulic monitoring; decision support; case studies

**Citation:** Limongelli, M.P.; Gentile, C.; Biondini, F.; di Prisco, M.; Ballio, F.; Belloli, M. The Monitoring Guidelines of the Lombardia Region in Italy. *Eng. Proc.* **2022**, *17*, 16. https://doi.org/10.3390/engproc2022017016

Academic Editor: Joaquín Martínez-Sánchez

Published: 16 May 2022

**Publisher's Note:** MDPI stays neutral with regard to jurisdictional claims in published maps and institutional affiliations.

On 14 August 2018, 43 people died in the collapse of the Polcevera bridge in Italy. Beyond the human tragedy, this event reminded us all of the degrading state of critical transport infrastructure in the EU. A few months after the tragedy, Regione Lombardia signed a collaboration agreement with Politecnico di Milano to develop criteria for the informed management and planning of interventions, aimed to keep the regional asset at the required performance level. One of the outcomes of the project are regional monitoring guidelines and their application to nine pilot bridges. The document provides guidance for the design of monitoring systems as decision support tools for problems relevant to maintenance and emergency management, occasional safety assessment, and standardization. In the MoRe regional guidelines, the design process of a monitoring system is approached as a stepwise procedure that originates from the needs of the decision-maker and comprises: (a) preliminary investigation to acquire knowledge about the specific bridge and the deterioration process to monitor; (b) the identification of the indicators able to provide information about the structural performance (deterioration processes, actions, and environmental conditions); (c) the selection of the technical devices to manage (acquire, process, transmit, and store) the monitoring information. During the lecture, the content of the guidelines and the nine pilot monitoring systems installed as demonstrators will be illustrated.

**Author Contributions:** Conceptualization, all authors; methodology, M.P.L.; investigation, C.G., F.B. (Fabio Biondini), M.d.P., F.B. (Francesco Ballio) and M.B.; writing—original draft preparation, M.P.L.; writing—review and editing, M.P.L.; visualization; project administration, M.B.; funding acquisition, M.B. All authors have read and agreed to the published version of the manuscript.

**Funding:** This research received was funded by Regione Lombardia–Politecnico di Milano agreement, 2018–2020.

**Data Availability Statement:** Data are not publicly available.

**Conflicts of Interest:** The authors declare no conflict of interest.

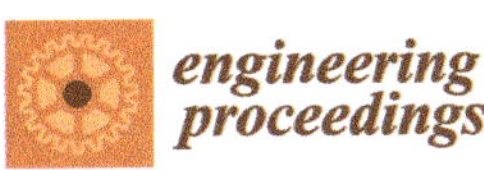

*engineering proceedings*

*Abstract*

# Integrated BIM-Based LCA for Road Asphalt Pavements †

**Salvatore Antonio Biancardo ***, **Cristina Oreto**, **Rosa Veropalumbo** and **Francesca Russo**

Department of Civil, Construction and Environmental Engineering, Federico II University of Naples, 80125 Naples, Italy; cristina.oreto@unina.it (C.O.); rosa.veropalumbo@unina.it (R.V.); francesca.russo2@unina.it (F.R.)

* Correspondence: salvatoreantonio.biancardo@unina.it; Tel.: +39-0817683772

† Presented at the 1st International Online Conference on Infrastructures, 7–9 June 2022; Available online: https://ioci2022.sciforum.net/.

**Keywords:** BIM; LCA; pavement maintenance

**Citation:** Biancardo, S.A.; Oreto, C.; Veropalumbo, R.; Russo, F. Integrated BIM-Based LCA for Road Asphalt Pavements. *Eng. Proc.* **2022**, *17*, 17. https://doi.org/10.3390/engproc2022017017

Academic Editor: Joaquín Martínez-Sánchez

Published: 16 May 2022

**Publisher's Note:** MDPI stays neutral with regard to jurisdictional claims in published maps and institutional affiliations.

In recent years, Building Information Modeling (BIM) tools have increased the productivity of infrastructure projects through more efficient information management and by fostering communication between different actors in the process [1–3]. At the same time, the growing need to introduce sustainability indicators, calculated through the life cycle assessment (LCA) methodology, has prompted an increase in the amount of data to be managed throughout the life cycle of an infrastructure project [4,5]. The present work consists of developing a BIM-based LCA tool aimed at the calculation of several environmental indicators through the informative content of a road pavement BIM; the tool is specifically designed to avoid errors in LCA calculations during the early design stages, reduce the engineer's effort through automation and support sustainable decision making in the infrastructure domain. An LCA-based pavement information model was developed by defining and adding several customized property sets, respectively, containing the specific road pavement materials' features and some selected environmental impact categories; a bidirectional information exchange path was established between BIM and the LCA tool to automate the LCA calculations and dynamically update the mentioned environmental indicators' property sets, whenever the geometry of the pavement and the asphalt materials' features change. The developed tool allows one to practically integrate pavement-related environmental sustainability requirements into BIM projects, with specific reference to asphalt pavement solutions that apply circular economy principles (i.e., secondary raw materials and cold recycling technologies), in light of more environmentally friendly pavement construction practices.

**Author Contributions:** Conceptualization, S.A.B. and F.R.; methodology, S.A.B. and F.R.; software, C.O. and R.V.; validation, S.A.B. and F.R.; data curation, C.O. and R.V.; visualization, C.O. and R.V.; supervision, S.A.B. and F.R. All authors have read and agreed to the published version of the manuscript.

**Funding:** This research received no external funding.

**Institutional Review Board Statement:** Not applicable.

**Informed Consent Statement:** Not applicable.

**Data Availability Statement:** The data presented in this study are available on request from the corresponding author.

**Conflicts of Interest:** The authors declare no conflict of interest.

## References

1.  Van Eldik, M.A.; Vahdatikhaki, F.; Dos Santos, J.M.O.; Visser, M.; Doree, A. BIM-based environmental impact assessment for infrastructure design projects. *Autom. Constr.* **2020**, *120*, 103379. [CrossRef]
2.  Bosurgi, G.; Pellegrino, O.; Sollazzo, G. Pavement condition information modelling in an I-BIM environment. *Int. J. Pavement. Eng.* **2021**. [CrossRef]
3.  Oreto, C.; Biancardo, S.A.; Veropalumbo, R.; Russo, F. Road Pavement Information Modeling through Maintenance Scenario Evaluation. *J. Adv. Transp.* **2021**, *2021*, 823117.
4.  Safari, K.; AzariJafari, H. Challenges and opportunities for integrating BIM and LCA: Methodological choices and framework development. *Sustain. Cities Soc.* **2021**, *67*, 102728. [CrossRef]
5.  Oreto, C.; Veropalumbo, R.; Viscione, N.; Biancardo, S.A.; Botte, M.; Russo, F. Integration of life cycle assessment into a decision support system for selecting sustainable road asphalt pavement mixtures prepared with waste. *Int. J. Life Cycle Assess.* **2021**, *26*, 2391–2407. [CrossRef]

*engineering proceedings*

**MDPI**

*Abstract*

# Pavement Information Modelling (PIM): Best Practice to Build a Digital Repository for Roads Asset Management [†]

**Orazio Baglieri [1,*], Anna Viola [1], Arianna Fonsati [2] and Anna Osello [2]**

[1]  Department of Environment, Land and Infrastructure Engineering (DIATI), Politecnico di Torino, 10129 Torino, Italy; violanna1990@gmail.com

[2]  Department of Structural, Building and Geotechnical Engineering (DISEG), Politecnico di Torino, 10129 Torino, Italy; arianna.fonsati@polito.it (A.F.); anna.osello@polito.it (A.O.)

*  Correspondence: orazio.baglieri@polito.it; Tel.: +39-0110905625

†  Presented at the 1st International Online Conference on Infrastructures, 7–9 June 2022; Available online: https://ioci2022.sciforum.net/.

**Keywords:** BIM; PIM; database

---

**Citation:** Baglieri, O.; Viola, A.; Fonsati, A.; Osello, A. Pavement Information Modelling (PIM): Best Practice to Build a Digital Repository for Roads Asset Management. *Eng. Proc.* **2022**, *17*, 18. https://doi.org/10.3390/engproc2022017018

Academic Editor: Cesare Sangiorgi

Published: 2 May 2022

**Publisher's Note:** MDPI stays neutral with regard to jurisdictional claims in published maps and institutional affiliations.

The application of BIM methods and tools plays a key role in transportation infrastructure asset management. Road pavements represent one of the main components of the asset, which greatly influences safety and quality of service for users. The work presented herein exploited the potentialities of BIM processes and methods for management of road pavement structures. The specific goal was to define best practice for development of a methodological framework for Pavement Information Modelling (PIM). The starting point of the process was the identification of the specific BIM use, as intended by Kreider and Messner [1]. In this case, the BIM use identified concerned the 3rd (3D), 4th (4D), and 5th (5D) dimensions of BIM. The adopted approach had the aim to define the steps to build PIM based on geometrical and structural parameters to be used as a database for different kinds of maintenance strategies. Within this context, the main objectives of the study can be summarized as follows:

(1)  Define the steps to develop a PIM including all the relevant information to be stored for management purposes, from data collection to data restitution,

(2)  Define a best practice for the integration among BIM tools and road pavement management methods in order to obtain a digital repository for predictive maintenance strategies,

(3)  Define a planning and cost database for the different technologies and materials involved in the different maintenance strategies.

From a practical point of view, the methodological framework was divided into three main categories (Figure 1) dealing with data: (i) data collection and input definition, which includes the analysis of available data and the BIM tools to be used to develop specific workflows; (ii) data processing, by dividing the workflows and related tasks in sub-sections for the fulfilment of the previously enounced objectives; (iii) data output, by defining the final result of each workflow.

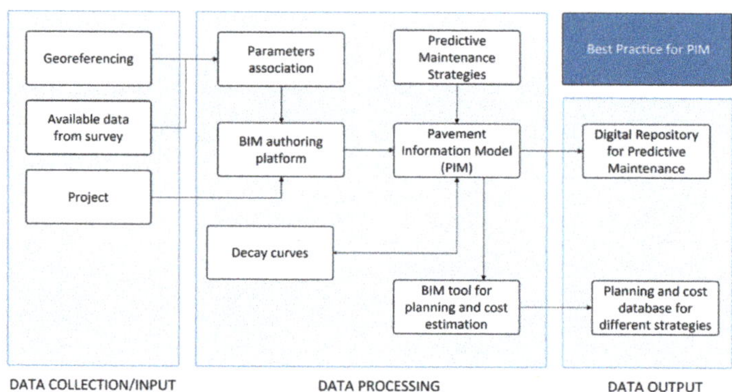

**Figure 1.** Pavement Information Modelling framework.

**Author Contributions:** Conceptualization, O.B., A.V. and A.O.; methodology, O.B., A.V. and A.F.; validation, A.F. and A.O.; formal analysis, A.V. and A.F.; investigation, O.B. and A.V.; resources, A.O.; data curation. A.V. and A.F.; writing—original draft preparation: A.V. and A.F.; writing—review and editing, O.B. and A.O.; visualization, A.V. and A.F.; supervision, O.B. and A.O. All authors have read and agreed to the published version of the manuscript.

**Funding:** This research received no external funding.

**Institutional Review Board Statement:** Not applicable.

**Informed Consent Statement:** Not applicable.

**Data Availability Statement:** Data sharing not applicable.

**Conflicts of Interest:** The authors declare no conflict of interest.

## Reference

1. Kreider, R.G.; Messner, J.I. *The Uses of BIM: Classifying and Selecting BIM Uses*; Version 0.9; The Pennsylvania State University: State College, PA, USA, 2013. Available online: http://bim.psu.edu (accessed on 1 February 2022).

*engineering proceedings*

**MDPI**

*Abstract*

# The Use of Tunnel Demolition Rocks to Produce Shotcrete for a Railway Infrastructure †

**Christian Paglia ***, **Giuliano Frigeri and Armando Chollet**

Institute of Materials and Construction, Supsi, V. Flora Ruchat 15, CH-6850 Mendrisio, Switzerland; giuliano.frigeri@supsi.ch (G.F.); armando.chollet@supsi.ch (A.C.)
* Correspondence: christian.paglia@supsi.ch
† Presented at the 1st International Online Conference on Infrastructures, 7–9 June 2022; Available online: https://ioci2022.sciforum.net/.

**Abstract:** Environmental issues are a main concern for society. The construction field largely affects the consumption of energy and the environment. In this concern, infrastructure, in particular streets, bridges and tunnels, is a necessity for life today. It links people, increasing their meeting capability. Tunnels have continuously gained relevant importance to shorten the travel distances and to save the landscape surface. In this work, demolished rock materials from the construction of part of a 50 km long tunnel through the alps were characterized and used to produce shotcrete to secure the tunnel walls. Several samples of demolished aggregates were investigated with respect to the granulometric curves. They needed to match with the reference curves in the content and amount of stone aggregates. This was particularly difficult in some cases because of the different mineralogy encountered. The type and form of the aggregates were also evaluated. These latter parameters have an influence on the workability and on the mechanical properties. In particular, the angular and subangular aggregates needed special attention. Then, the material was mixed by adding silica fume. This enabled a more dense microstructure by reducing the porosity at a later stage. The steel fibers were also added to the mixtures in different amounts to produce the shotcrete. The fresh concrete properties were measured directly on site. Furthermore, the hardened state was controlled on site and in the laboratory. The compression strength exhibited variable values, which could be related to the mixing proportion of the ingredients. The punch tests indicated similar fracture behaviors but were very important for the safety of the worker inside the tunnel, in particular where material enrichment was present on the roofing parts. The steel fiber content generally increased the ductility of the specimens. The porosity and the water permeability were controlled, as well as the freeze/thaw resistance. The mixtures were continuously optimized by keeping the water/cement ratio and the superplasticizer dosage under control. All these adaptations allowed for the reuse of a large amount of the tunnel demolition material. The concrete was produced in a special mixing plant on site. This reduced the transportation and increased the environmental sustainability of such a long infrastructure.

**Keywords:** tunnel; demolition; rock; aggregates; shotcrete

**Citation:** Paglia, C.; Frigeri, G.; Chollet, A. The Use of Tunnel Demolition Rocks to Produce Shotcrete for a Railway Infrastructure. *Eng. Proc.* **2022**, *17*, 19. https://doi.org/10.3390/engproc2022017019

Academic Editor: Luigi Coppola

Published: 2 May 2022



**Author Contributions:** C.P.: abstract and presentation preparation and talk, Data gathering and evaluation, G.F. and A.C.: field and laboratory measurements, data acquisition and evaluation. All authors have read and agreed to the published version of the manuscript.

**Funding:** Public/private collaboration internal mandate 4400.

**Institutional Review Board Statement:** Not applicable.

**Informed Consent Statement:** Not applicable.

**Data Availability Statement:** Archive of the Institute of materials and construction.

**Conflicts of Interest:** The authors declare no conflict of interest.

*engineering proceedings*

*Abstract*

# Investigating the Viability of Multi-Recycling of Asphalt Mixtures through a Preliminary Binder Level Characterization [†]

Gaetano Di Mino, Konstantinos Mantalovas and Vineesh Vijayan *

Department of Engineering, University of Palermo, 90128 Palermo, Italy; gaetano.dimino@unipa.it (G.D.M.); konstantinos.mantalovas@unipa.it (K.M.)
* Correspondence: vineesh.vijayan@unipa.it
† Presented at the 1st International Online Conference on Infrastructures, 7–9 June 2022; Available online: https://ioci2022.sciforum.net/.

**Abstract:** The incorporation of reclaimed asphalt (RA) in hot mix asphalt mixtures is widely considered a sustainable solution for road infrastructure development. Under the scope of the circular economy (CE), the multiple recycling capability of RA has to be assessed in order to ensure its performance at each recycling cycle and also its viability with different additives. The performance of asphalt mixtures with RA strongly depends on the type of rejuvenator, binder, and their degree of blending in the mix. For this reason, it is essential to know the properties of the aged binder extracted from RA to better understand its rheological properties and optimal dosage of rejuvenation to design a satisfactory blend design for the recycled mixture. To analyse the multi-recycling potential of the recycled mixture with high RA content, it is imperative to study its characteristics at every recycling cycle. Therefore, in this study, a preliminary binder-scale study is carried out to better understand the ageing, rejuvenating effects and morphological changes that occur on the bituminous binders at every recycling cycle. The study has been conducted on a RA binder, extracted from RA from a rural road in Italy and the simulation of multiple recycling is conducted through a laboratory ageing protocol on both binder and asphalt mixture scales. The long-term binder level ageing is performed by a pressure ageing vessel (PAV) after the short-term ageing by the rolling thin film oven test (RTFO). The asphalt mixture ageing is performed through a protocol similar to the National Cooperative Highway Research Program (NCHRP) and the aged binder is extracted from the mixture for further investigations. Multiple recycling is simulated by repeating the ageing procedure after rejuvenating both the aged binder and aged mixture up to the amount of recycling needed for the study. The rheological properties of the aged binder obtained from both binder-scale and mixture-scale ageing methods are evaluated using a dynamic shear rheometer (DSR) and bending beam rheometer (BBR). Moreover, the morphological changes that occurred are analysed using SARA (saturates, aromatics, resin and asphaltenes) fractionation and atomic force microscopy (AFM). The results of the study can help towards answering the uncertainties regarding the performance of high RA% in asphalt mixtures and establishing its viability in multi recycling towards the full-scale implementation of this sustainable approach.

**Keywords:** multi-recycling; rejuvenators; reclaimed asphalt binders; artificial ageing; aged binder; circular economy; binder morphology

**Citation:** Di Mino, G.; Mantalovas, K.; Vijayan, V. Investigating the Viability of Multi-Recycling of Asphalt Mixtures through a Preliminary Binder Level Characterization. *Eng. Proc.* **2022**, *17*, 20. https://doi.org/10.3390/engproc2022017020

Academic Editor: Andrea Graziani

Published: 2 May 2022

**Publisher's Note:** MDPI stays neutral with regard to jurisdictional claims in published maps and institutional affiliations.

**Author Contributions:** Conceptualization, G.D.M., K.M. and V.V.; Methodology, G.D.M., V.V.; Validation, G.D.M., K.M. and V.V.; Formal Analysis, V.V.; Investigation, V.V.; Data curation, V.V.; Writing—Original Draft Preparation, G.D.M., K.M. and V.V.; Writing—Review and Editing, G.D.M., K.M. and V.V.; Visualization, V.V.; Supervision, G.D.M. and K.M.; Project Administration, G.D.M. All authors have read and agreed to the published version of the manuscript.

**Funding:** This research received no external funding.

**Conflicts of Interest:** The authors declare no conflict of interest.

*engineering proceedings*

*Abstract*

# Introduction to a New Extrusion-Based Technology for the Regeneration of Existing Tunnels [†]

Andrea Marcucci [1,*], Stefano Guanziroli [2], Alberto Negrini [2], Liberato Ferrara [1] and Bernardino Chiaia [3]

[1] Department of Civil and Environmental Engineering, Politecnico di Milano, 20133 Milan, Italy; liberato.ferrara@polimi.it

[2] Hinfra, 15033 Casale Monferrato, Italy; stefano.guanziroli@hinfra.it (S.G.); alberto.negrini@hinfra.it (A.N.)

[3] Department of Structural Geotechnical and Building Engineering, Politecnico di Torino, 10129 Turin, Italy; bernardino.chiaia@polito.it

* Correspondence: andrea.marcucci@polimi.it

† Presented at the 1st International Online Conference on Infrastructures, 7–9 June 2022; Available online: https://ioci2022.sciforum.net/.

**Keywords:** additive manufacturing; 3D concrete printing; slip-forming; tunnels; fiber reinforced concrete

**Citation:** Marcucci, A.; Guanziroli, S.; Negrini, A.; Ferrara, L.; Chiaia, B. Introduction to a New Extrusion-Based Technology for the Regeneration of Existing Tunnels. *Eng. Proc.* **2022**, *17*, 21. https://doi.org/10.3390/engproc2022017021

Academic Editor: Piergiorgio Tataranni

Published: 2 May 2022

**Publisher's Note:** MDPI stays neutral with regard to jurisdictional claims in published maps and institutional affiliations.

Additive Manufacturing (AM) is process in which a three-dimensional component is produced by the consecutive addition of material. This technology, applied on a large scale to cementitious materials, is known as 3D Concrete Printing (3DCP). Among the new technologies driving the fourth industrial revolution in the construction industry, 3D Concrete Printing (3DCP) is playing a key role. The typical process is carried out through robotic arms or gantries equipped with nozzles, similarly to contour crafting in other industries, where the printed object is obtained through the multiple deposition of layers. Although 3DCP is appealing when applied to specific items, as complex architectural shapes, the structural behavior and geometrical size are limitations that are difficult to overcome. Upscaling the extrusion process to full scale infrastructure applications through introducing a new concept of ultrafast and adaptable slip forming is the key to accessing different domains of the industry, where the increase in productivity results in social, economic and environmental benefits that are not comparable to the niche to which 3DCP is confined. As a matter of fact, the process of maintaining existing infrastructures is a very critical topic in most of the industrialized countries worldwide. It is commonly recognized by the main players operating in the industry (professional engineers, owners, construction companies, etc.) that, despite for new constructions, the methodologies are quite evolved (i.e., development of the tunnel boring machines), in the maintenance area there is complete lack of technologies, making it still impossible to industrialize the operations. This paper will present the Extruded Tunnel Lining Regeneration (ETLR) technology developed by HINFRA, which can automatically regenerate the lining of existing damaged tunnels directly at the site. The ETRL processing train is a machine consisting of several modular units, each solving a specific function. The increasing industrialization of operations, typically the demolition of the existing lining, the surface preparation and the new lining phases, combined with the performances of the advanced concrete, allow for targeting better productivity rates than those achieved with the traditional methods in the industry. This is made possible by the development of an extrudable eco-friendly Fiber-Reinforced Concrete (FRC) characterized by high early-age compressive strength and a fast setting time, which is the other key aspect of the innovative technology implemented by HINFRA. "Tailored" technological issues, including, e.g., the experimental determination of the friction between the extrudable mixes and formworks, will be discussed, together with a design validation related to a FRC tunnel lining, whose use could further exploit, through the significant reduction in ordinary reinforcement, the potential of 3DCP.

**Author Contributions:** Conceptualization, S.G., L.F. and B.C.; methodology, L.F., A.M. and S.G.; software, A.M.; validation, S.G., L.F. and A.M.; formal analysis, A.M.; investigation, A.M.; resources, S.G., L.F. and B.C.; data curation, A.M. and L.F.; writing—original draft preparation, A.M. and A.N.; writing—review and editing, A.M., A.N., S.G. and L.F.; visualization, A.M.; supervision, S.G. and L.F. All authors have read and agreed to the published version of the manuscript.

**Funding:** The first author acknowledges the support of the Italian National Programme PON Ricerca e Innovazione in funding his PhD programme.

**Institutional Review Board Statement:** Not applicable.

**Informed Consent Statement:** Not applicable.

**Data Availability Statement:** Data presented in this research are original and all produced by the authors.

**Conflicts of Interest:** The authors declare no conflict of intertest.

*engineering proceedings*

*Abstract*

# Lifecycle Assessment of Permeable Interlocking Concrete Pavement and Comparison with Conventional Mixes [†]

Sandhya Makineni [1], Avishreshth Singh [2],[*] and Krishna Prapoorna Biligiri [1]

[1] Department of Civil & Environmental Engineering, Indian Institute of Technology, Tirupati 517619, India; ce22d002@iittp.ac.in (S.M.); bkp@iittp.ac.in (K.P.B.)

[2] Civil Engineering Department, Indian Institute of Technology, Tirupati 517619, India

[*] Correspondence: avi.theriac@gmail.com; Tel.: +91-9997847652

[†] Presented at the 1st International Online Conference on Infrastructures, 7–9 June 2022; Available online: https://ioci2022.sciforum.net/.

**Keywords:** pervious interlocking concrete pavement; cement concrete; asphalt concrete; parking lot; lifecycle assessment; pavements; sustainable infrastructure

**Citation:** Makineni, S.; Singh, A.; Biligiri, K.P. Lifecycle Assessment of Permeable Interlocking Concrete Pavement and Comparison with Conventional Mixes. *Eng. Proc.* **2022**, *17*, 22. https://doi.org/10.3390/engproc2022017022

Academic Editor: Ana Jiménez Del Barco Carrión

Published: 2 May 2022

**Publisher's Note:** MDPI stays neutral with regard to jurisdictional claims in published maps and institutional affiliations.

In recent years, continuous attempts have been made by the pavement industry to explore the opportunities that assist in bringing down the environmental footprint of roadway infrastructure as well as mitigate the harmful impacts of climate change on the quality-of-life. The construction of pervious interlocking concrete pavement (PICP) in parking areas is gaining widespread acceptance attributed to their: (a) ease of installation, (b) high durability and skid resistance, (c) low repair and maintenance requirements, (d) ability to mitigate floods, and (e) potential to purify stormwater. However, very little research has been conducted to investigate the environmental impacts associated with the installation of such pavement systems. Therefore, the objective of this cradle-to-gate research study was to quantify the environmental footprint of PICP for a 75 m × 16.5 m parking lot that was constructed in the premises of the Indian Institute of Technology Tirupati, India. Further, the quantified impacts were compared to that of traditional asphalt concrete (AC) and cement concrete (CC) parking lots. The scope of the effort encompassed: (a) design of three pavement systems based on site specific requirements as per relevant design codebooks, and (b) quantification of the environmental impacts using systematic lifecycle assessment (LCA) approaches that are in accordance with the international standards. The results indicated that construction of an AC parking lot had a lower environmental footprint compared to CC pavement and PICP systems. Further, the environmental impacts associated with the construction of CC pavements were the highest. Based on the results, it was understood that though the PICP system has an intermediate environmental footprint, it provides additional benefits such as infiltration of stormwater into the ground. Further, the PICP blocks have higher design life compared to CC and AC pavements. However, additional research must be conducted in the future to ascertain the environmental impacts of the three pavement systems from a cradle-to-grave perspective. Such an approach will assist in the integration of LCA toolkits with existing pavement design methods, and further contribute to the development of resilient and sustainable pavement infrastructure.

**Author Contributions:** Conceptualization, S.M., A.S. and K.P.B.; methodology, S.M., A.S.; software, S.M. and A.S.; validation, S.M. and A.S.; formal analysis, S.M. and A.S.; investigation, S.M. and A.S.; resources, S.M. and A.S.; data curation, S.M. and A.S.; writing—original, S.M., A.S. and K.P.B.; draft preparation, S.M., A.S. and K.P.B.; writing—review, A.S. and K.P.B.; editing, A.S. and K.P.B.; visualization, S.M., A.S. and K.P.B.; supervision, A.S. and K.P.B.; project administration, A.S. and K.P.B.; funding acquisition, K.P.B. All authors have read and agreed to the published version of the manuscript.

**Funding:** This research received no external funding.

**Institutional Review Board Statement:** Not applicable.

**Informed Consent Statement:** Not applicable.

**Data Availability Statement:** Not applicable.

**Acknowledgments:** The authors are thankful to the Indian Institute of Technology Tirupati, India, for providing access to the necessary software required for completion of this study.

**Conflicts of Interest:** The authors declare no conflict of interest.

*Abstract*

# News Applications of UAVs for Infrastructure Monitoring: Contact Inspection Systems †

**L. M. González-deSantos** 

Engineering Physics Group, School of Aerospace Engineering, Campus Ourense, University of Vigo, 32004 Ourense, Spain; luismgonzalez@uvigo.es

† Presented at the 1st International Online Conference on Infrastructures, 7–9 June 2022; Available online: https://ioci2022.sciforum.net/.

**Keywords:** NDT inspections; contact inspections; UAV payload; SHM



**Citation:** González-deSantos, L.M. News Applications of UAVs for Infrastructure Monitoring: Contact Inspection Systems. *Eng. Proc.* **2022**, *17*, 23. https://doi.org/10.3390/engproc2022017023

Academic Editor: Joaquín Martínez-Sánchez

Published: 6 May 2022

**Publisher's Note:** MDPI stays neutral with regard to jurisdictional claims in published maps and institutional affiliations.



In recent years, the use of UAVs (Unmanned Aerial Vehicles), as known as drones, has increased exponentially for infrastructure monitoring, usually using remote sensing payloads. The drop in prices of these systems, the improvements in their specifications, and the change in the regulations for their use have given more and more people access to use them for both recreational and professional means. In some hard-to-access structures, such as bridges or dams, these vehicles are a powerful tool to carry out different types of inspections using remote sensors, such as different types of camaras, LiDAR sensors, or RADAR sensors. The data acquired by these vehicles can be used by SHM (Structural Health Monitoring) methods to acquire a 3D geometric model of the structure to be used by a DT (digital twin) or to detect different pathologies, such as cracks. Additionally, new UAV systems have been developed in recent years to perform a physical contact between the UAV and the structure, enabling the use of these systems to perform other NDT (Non-Destructive Testing) inspections that use sensors that have to be in contact with the structure to perform reliable measurements, such as ultrasonic sensors. In this work, four different intelligent payloads for contact inspection tasks with UAVs are going to be presented. The first three payloads [1–3] are focused on maintaining continuous contact between the UAV and the structure while measurements are performed by the contact sensor. Instead, the fourth has been designed to fix the payload to the structure. In this way, the UAV only fixes it to the structure without maintaining continuous contact while the measurements are performed. The results of each payload are going to be compared and analysed, defining possible improvements and future work.

**Funding:** This study was funded by the Recovery, Transformation, and Resilience Plan of the European Union–Next Generation EU (University of Vigo grant ref. 585507).

**Informed Consent Statement:** Not applicable.

**Conflicts of Interest:** The authors declare no conflict of interest.

## References

1. González-deSantos, L.M.; Martínez-Sánchez, J.; González-Jorge, H.; Ribeiro, M.; de Sousa, J.B.; Arias, P. Payload for Contact Inspection Tasks with UAV Systems. *Sensors* **2019**, *19*, 3752. [CrossRef]
2. González-deSantos, L.M.; Martínez-Sánchez, J.; González-Jorge, H.; Navarro-Medina, F.; Arias, P. UAV payload with collision mitigation for contact inspection. *Autom. Constr.* **2020**, *115*, 103200. [CrossRef]
3. González-deSantos, L.M.; Martínez-Sánchez, J.; González-Jorge, H.; Arias, P. Active UAV payload based on horizontal propellers for contact inspections tasks. *Meas. J. Int. Meas. Confed.* **2020**, *165*, 108106. [CrossRef]

*Abstract*

# High-Temperature, Bond, and Environmental Impact Assessment of Alkali-Activated Concrete (AAC) †

**Kruthi Kiran Ramagiri** [1,*] , **Patricia Kara De Maeijer** [2] and **Arkamitra Kar** [1]

1 Department of Civil Engineering, Birla Institute of Technology and Science-Pilani, Hyderabad Campus, Hyderabad 500078, India; arkamitra.kar@hyderabad.bits-pilani.ac.in
2 Built Environment Assessing Sustainability (BEASt), EMIB, Faculty of Applied Engineering, University of Antwerp, Groenenborgerlaan 171, 2020 Antwerp, Belgium; patricija.karademaeijer@uantwerpen.be
* Correspondence: p20170008@hyderabad.bits-pilani.ac.in
† Presented at the 1st International Online Conference on Infrastructures, 7–9 June 2022; Available online: https://ioci2022.sciforum.net/.

**Keywords:** alkali-activated binder (AAB); alkali-activated concrete (AAC); high temperature; sustainability; microstructure; life cycle assessment (LCA)

**Citation:** Ramagiri, K.K.; Kara De Maeijer, P.; Kar, A. High-Temperature, Bond, and Environmental Impact Assessment of Alkali-Activated Concrete (AAC). *Eng. Proc.* **2022**, *17*, 24. https://doi.org/10.3390/engproc2022017024

Academic Editor: Aljoša Šajna

Published: 2 May 2022

**Publisher's Note:** MDPI stays neutral with regard to jurisdictional claims in published maps and institutional affiliations.

Alkali-activated binder (AAB) has been extensively researched in recent years due to its potential to replace Portland cement (PC) and lower carbon footprint. However, major barriers to its commercialization are related to the inadequate characterization of mechanical properties and long-term durability. The mechanical and durability performance of AAB is highly influenced by its microstructure. There is minimal research on correlating the microstructural changes to the specimen-level performance of AAB [1]. Among AAB's primary advantages as a building material is its superior performance at high temperatures and lower environmental impact [2]. The performance of reinforced concrete to function as a composite at high temperatures is evaluated through its bond strength. Several studies reported the effect of mix proportions, curing conditions, and rebar specifications on the bond strength of thermal-cured alkali-activated concrete (AAC) [3–6]. However, there is no reported study on the bond strength of ambient cured (fly ash + slag)-based AAC. To validate the practical sustainability of AAC, life cycle assessment (LCA) can be used to evaluate the environmental impact.

Therefore, the present study evaluates the effect of varying precursor proportion (fly ash: slag varied as 100:0, 70:30, 60:40, and 50:50), activator modulus (Ms, varied as 1.0 and 1.4), and high temperatures (538 °C, 760 °C, and 892 °C) on the mechanical properties and microstructure of AAC. The microstructural characteristics are evaluated using X-ray diffraction (XRD), Fourier transform infrared spectroscopy (FTIR), and scanning electron microscopy coupled with energy-dispersive X-ray spectroscopy (SEM-EDS). The effect of varying precursor proportions and Ms on the mechanical performance of AAC is evaluated through compressive strength, bond strength, flexural strength, and split tensile strength testing. The performance of AAB at extremely high temperatures is assessed in terms of residual compressive and bond strength. LCA of AAC is conducted using the ReCiPe 2016 methodology. Furthermore, since the commercialization of any novel alternative material depends on cost-effectiveness, a simplified cost analysis is performed.

The results from microstructural experiments show the formation of new crystalline phases and decomposition of reaction products when exposed to high temperatures, and they correlate well with the observed mechanical performance. The 28-day compressive strength with slag content is enhanced by 151.8–339.7%, depending on the mix. In ambient conditions, lower Ms improves mechanical performance. When exposed to high temperatures, specimens with a high slag content and a low Ms suffered significant deterioration. AAC with a fly ash: slag ratio of 70:30 and Ms of 1.4 is proposed as optimal from the results obtained in the present study [7]. The results reveal that the biggest impact on climate change comes from transport (45.5–48.2%) and sodium silicate (26.7–35.6%). Environmental impact is determined to be primarily influenced by sodium hydroxide. The proposed optimal AAC mix has a global warming potential 42.6 % lower than PC concrete [8]. A comparison with the default procedures in the International Reference Life Cycle Data System (ILCD) handbook reveals that the ReCiPe midpoint approach is more efficient in analyzing all impact categories, except freshwater ecotoxicity (FETP) and human toxicity potentials (HTPs). An evaluation of FETP and HTP is recommended with USEtox [9]. The proposed AAC mix has a higher cost than PC concrete in the present scenario. In contrast, if a carbon tax is enacted, the cost of the proposed AAC mix will rise by only 18.4%, whereas PC concrete prices will rise by 81.7%. This proposed AAC mix is an environmentally sustainable replacement for PC concrete specifically intended for applications requiring the superior high-temperature performance of reinforced concrete.

**Author Contributions:** Conceptualization, K.K.R. and A.K.; methodology, K.K.R.; writing—original draft preparation, K.K.R.; writing—review and editing, A.K. and P.K.D.M.; supervision, A.K.; funding acquisition, A.K. All authors have read and agreed to the published version of the manuscript.

**Funding:** This research was funded by BITS Pilani, Hyderabad campus, through the Outstanding Potential for Excellence in Research and Academics (OPERA) grant.

**Institutional Review Board Statement:** This study does not involve any studies on humans or animals.

**Informed Consent Statement:** Not applicable.

**Acknowledgments:** The authors would like to acknowledge the central analytical laboratory facilities at BITS Pilani, Hyderabad campus, for providing the necessary setup to conduct XRD, FTIR, and SEM-EDS analyses.

**Conflicts of Interest:** The authors declare no conflict of interest.

## References

1. Kar, A. Characterizations of Concretes with Alkali-Activated Binder and Correlating Their Properties from Micro-to Specimen Level. Ph.D. Thesis, West Virginia University, Morgantown, WV, USA, 2013.
2. Ramagiri, K.K. Evaluation of High-Temperature, Bond, and Shrinkage of Alkali-Activated Binder Concrete. Ph.D. Thesis, Birla Institute of Technology and Science, Pilani, India, 2021.
3. Adak, D.; Sarkar, M.; Mandal, S. Structural performance of nano-silica modified fly ash based geopolymer concrete. *Constr. Build. Mater.* **2017**, *135*, 430–439. [CrossRef]
4. Sarker, P.K. Bond strength of reinforcing steel embedded in fly ash-based geopolymer concrete. *Mater. Struct.* **2011**, *44*, 1021–1030. [CrossRef]
5. Castel, A.; Foster, S.J. Bond strength between blended slag and Class F fly ash geopolymer concrete with steel reinforcement. *Cem. Concr. Res.* **2015**, *72*, 48–53. [CrossRef]
6. Sofi, M.; van Deventer, J.S.J.; Mendis, P.A.; Lukey, G.C. Bond performance of reinforcing bars in inorganic polymer concrete (IPC). *J. Mater. Sci.* **2007**, *42*, 3107–3116. [CrossRef]
7. Ramagiri, K.K.; Chauhan, D.R.; Gupta, S.; Kar, A.; Adak, D.; Mukherjee, A. High-temperature performance of ambient-cured alkali-activated binder concrete. *Innov. Infrastruct Solut.* **2021**, *6*, 1–11. [CrossRef]
8. Ramagiri, K.K.; Chintha, R.; Bandlamudi, R.K.; Kara De Maeijer, P.; Kar, A. Cradle-to-gate life cycle and economic assessment of sustainable concrete mixes—alkali-activated concrete (AAC) and bacterial concrete (BC). *Infrastructures* **2021**, *6*, 104. [CrossRef]
9. Ramagiri, K.K.; Kar, A. Environmental impact assessment of alkali-activated mortar with waste precursors and activators. *J. Build. Eng.* **2021**, *44*, 103391. [CrossRef]

*engineering proceedings*

*Abstract*

# Increasing the Use of Reclaimed Asphalt in Italy towards a Circular Economy: A Top-Down Approach [†]

Iain Peter Dunn [1,*], Francesco Acuto [1], Oumaya Yazoghli-Marzouk [2], Gaetano Di Mino [1] and Konstantinos Mantalovas [1]

1 Department of Engineering (DIING), University of Palermo, 90128 Palermo, Italy; francesco.acuto@unipa.it (F.A.); gaetano.dimino@unipa.it (G.D.M.); konstantinos.mantalovas@unipa.it (K.M.)
2 Cerema, Direction Centre Est, Agence Autun, Boulevard Giberstein, BP 141, F-71405 Autun, France; oumaya.marzouk@cerema.fr
* Correspondence: iainpeter.dunn@unipa.it
† Presented at the 1st International Online Conference on Infrastructures, 7–9 June 2022; Available online: https://ioci2022.sciforum.net/.

**Keywords:** reclaimed asphalt; circular economy; sustainability; recycling; legislations

**Citation:** Dunn, I.P.; Acuto, F.; Yazoghli-Marzouk, O.; Di Mino, G.; Mantalovas, K. Increasing the Use of Reclaimed Asphalt in Italy towards a Circular Economy: A Top-Down Approach. *Eng. Proc.* **2022**, *17*, 25. https://doi.org/10.3390/engproc2022017025

Published: 2 May 2022

**Publisher's Note:** MDPI stays neutral with regard to jurisdictional claims in published maps and institutional affiliations.

This contribution concerns recommendations which could be made to Italian regulatory bodies to improve their use of reclaimed asphalt (RA) in the road engineering sector. It is essential for the use of RA to be established as a standard practice since it has been proven that it can serve as an end-of-waste product that complies with the principles of a circular economy within both an open- and a closed-loop approach [1,2]. Several aspects will be covered, starting with an analysis of European nations whose economies can be classed as "more circular" compared to Italy's. This refers to nations which have a high usage of RA. Furthermore, nations which have clear regulatory guidelines on the use of RA in road construction or, alternatively, nations which have very lax statutory requirements on pavement design allowing best practice to reign could be considered as more circular if these regulations or permissiveness result in a greater uptake in the use of RA in road construction.

The European average of RA reused in pavement construction currently lies at 60% [3]; however, some nations greatly outperform this average. Some examples of nations which have the highest usage of RA in pavements are Germany, France, and Spain, which report using 84%, 76%, and 72.7% of all reclaimed asphalt in pavement activities, respectively [4]. To understand why this is possible in these nations which greatly outperform Italy, which, according to the same source, reuses only 25% of available RA [4], it is necessary to understand the regulatory framework in each nation, promoting the use of RA in pavement design while limiting RA's use in Italy.

The most interesting example listed above, in the context of making recommendations to an Italian regulatory body, may be Spain; this is a nation with a similar economic capacity and a similar climatic condition to Italy that would permit a similar use of RA, prompting the question, why is there such a vast chasm between Spanish and Italian figures on the use of RA in road construction?

The answer, in the authors' opinion, is that Spain sees the use of RA in pavement management not only as a sustainable solution but also as one that is cost-effective. Moreover, Spain appears to be an early adopter of hot and warm mix asphalt recycling, beginning in the 1980s [5]. These facts, combined with a wealth of experience gained over the past three decades, seem to have changed the perspective of Spanish lawmakers, who now understand that pavements containing a percentage of RA are not inferior to those containing only virgin materials [5]. Additionally, there is a dearth of research supporting the use of RA and advocating that a higher percentage of RA be used in pavement design.

Combining the use of rejuvenators, asphalt mixtures with an RA content of 40% allow not only the amount of virgin aggregates but also the quantity of virgin binders to be reduced [6]. Assuming that Spain is as receptive to this new research as it has been in the past, it is likely that its rate of reuse of RA will continue to increase.

In Italy, however, there seems to be little push to increase the use of RA in road construction. The Italian association of pavement design and bitumen—SITEB—cites several obstacles: firstly, complex bureaucracy and the slow rate of change to regulations; secondly, non-uniform regulations which vary not only from region to region but also from municipality to municipality [7]; and lastly, a prejudice among not only engineers but also road authorities and governmental bodies against the use of RA. Moreover, the Italian regulatory context allows for only 30%, 25%, and 20% of RA usage in bases, binders, and surface courses, a fact that significantly limits and hinders the exploitation of RA as an end-of-waste product. On the contrary, in Spain, although mixtures composed of 60–70% recycled materials can be produced, the most common practice is the production of asphalt mixtures with an RA content below 50%. Thus, it becomes evident that the increase in the allowed RA% in the recycling process of asphalt mixtures can significantly impact the recycling and sustainability implications of a country.

According to a report dating from 2011, Italy had the second highest quantity of available RA which could be used in new construction, and yet only 20% of that material was used [8]. At that point in time, Germany had the highest production and the highest recycling rate (82%) of any European nation. It is disheartening to see that the improvement in Italy's use of RA has been slow, and the authors would like to make recommendations to Italian regulatory bodies, in the hope that they could learn not only from their most adept and advanced in terms of RA recycling European partners but also from those in a similar economic condition who understand that the use of RA is not only environmentally sustainable but also economically sustainable.

**Author Contributions:** Conceptualization, K.M. and I.P.D.; Methodology, K.M. and I.P.D.; Validation, K.M. and I.P.D.; Formal Analysis, K.M. and I.P.D.; Investigation, K.M. and I.P.D.; Data curation, K.M. and I.P.D.; Writing—Original Draft Preparation, K.M. and I.P.D.; Writing—Review and Editing, K.M., I.P.D., F.A., O.Y.-M. and G.D.M.; Supervision, G.D.M. Project Administration, G.D.M. All authors have read and agreed to the published version of the manuscript.

**Funding:** This project has received funding from the European Union, ENI CBC Med Program, Education, Research, Technological Development and Innovation, under grant agreement n°28/1682.

**Conflicts of Interest:** The authors declare no conflict of interest.

## References

1. Mantalovas, K.; di Mino, G. The Sustainability of Reclaimed Asphalt as a Resource for Road Pavement Management through a Circular Economic Model. *Sustainability* **2019**, *11*, 2234. [CrossRef]
2. Mantalovas, K.; di Mino, G. Integrating Circularity in the Sustainability Assessment of Asphalt Mixtures. *Sustainability* **2020**, *12*, 594. [CrossRef]
3. Jaawani, S.; Franco, A.; de Luca, G.; Coppola, O.; Bonati, A. Limitations on the Use of Recycled Asphalt Pavement in Structural Concrete. *Appl. Sci.* **2021**, *11*, 10901. [CrossRef]
4. EUROPEAN ASPHALT PAVEMENT ASSOCIATION. EAPA—Asphalt in Figures 2020. Available online: https://096.wpcdnnode.com/eapa.org/wp-content/uploads/2021/12/asphalt_in_figures_2020.pdf (accessed on 2 February 2022).
5. Martinez-Echevarria, M.J.; Rubio, M.C.; Menéndez, A. The Reuse Of Waste From Road Resurfacing: ColdIn-place Recycling Of Bituminous Pavement, An Environmentally Friendly Alternative To Conventional Pavement Rehabilitation Methods. *Waste Manag.* **2008**, *109*, 459–469.
6. Rodríguez-Fernández, I.; Lastra-González, P.; Indacoechea-Vega, I.; Castro-Fresno, D. Recyclability potential of asphalt mixes containing reclaimed asphalt pavement and industrial by-products. *Environ. Sci. Constr. Build. Mater.* **2019**, *195*, 148–155. [CrossRef]
7. Strade: In Italia, Solo il 25% del Fresato D'asfalto Viene Avviato al Recupero. Available online: https://www.ingenio-web.it/26047-strade-in-italia-solo-il-25-del-fresato-dasfalto-viene-avviato-al-recupero (accessed on 2 February 2022).
8. Ravaioli, S.; SITEB. Reclaimed Asphalt: Waste or By-Product? 2011. Available online: https://www.siteb.it/wp-content/uploads/rassegna_del_bitume/articoli/6911_2.pdf (accessed on 2 February 2022).

**engineering proceedings**

*Abstract*

# Bituminous Interlayers Thermomechanical Behaviour under Small Shear Strain Loading Cycles with 2T3C Apparatus: Hollow Cylinder and Digital Image Correlation [†]

**Thien Nhan Tran *, Salvatore Mangiafico , Cédric Sauzéat and Hervé Di Benedetto**

Univ Lyon, ENTPE, Ecole Centrale de Lyon, CNRS, LTDS, UMR5513, 69518 Vaulx en Velin, France;
salvatore.mangiafico@entpe.fr (S.M.); cedric.sauzeat@entpe.fr (C.S.); herve.dibenedetto@entpe.fr (H.D.B.)
* Correspondence: thiennhan.tran@entpe.fr
† Presented at the 1st International Online Conference on Infrastructures, 7–9 June 2022;
   Available online: https://ioci2022.sciforum.net/.

**Keywords:** bituminous mixture; interface; shear loading; DIC technology

**Citation:** Tran, T.N.; Mangiafico, S.; Sauzéat, C.; Di Benedetto, H. Bituminous Interlayers Thermomechanical Behaviour under Small Shear Strain Loading Cycles with 2T3C Apparatus: Hollow Cylinder and Digital Image Correlation. *Eng. Proc.* **2022**, *17*, 26. https://doi.org/10.3390/engproc2022017026

Academic Editor: Augusto Cannone Falchetto

Published: 2 May 2022

**Publisher's Note:** MDPI stays neutral with regard to jurisdictional claims in published maps and institutional affiliations.

## 1. Introduction

A road/airport pavement is a multi-layered structure generally composed of several layers of bituminous materials, and cement-bound materials on unbound granular materials. In the design phase, the different bituminous layers are considered perfectly bonded and therefore expected to work as a unique structure throughout the service life of the pavement. However, due to environmental, traffic, and/or material-related conditions, the quality of the bond changes over time. The layers in the structure tend to work more and more independently with a degrading bonding capacity, which induces a reduction in the pavement life. Therefore, the mechanical behavior of the interface between bituminous layers has recently been studied through some original approaches. There were tests developed to focus on the interface study, for example, the Ancona shear testing research and analysis device at the Marche Polytechnic University [1–3], the shear-torque fatigue testing device at the University of Limoges [4,5], and other tests [6]. Most of the studies on this topic present several limitations. Only one or a few loading configurations can be applied to the sample (for example, pure shear of the interface) and studies focus only on the interface strength. Moreover, stress and strain fields within the sample are not homogenous, therefore not allowing investigation of the intrinsic mechanical behavior of the interface. In this study, the 2T3C ("Torsion, Traction/Compression sur Cylindre Creux" in French, Torsion, Traction/Compression on Hollow Cylinder) apparatus developed at ENTPE is used to investigate the behavior of a bituminous interface under shear loading and small strain cycles.

## 2. T3C Apparatus

The device consists of different basic parts: (1) a servo-hydraulic press capable of imposing axial and shear loading (cyclic or monotonic) on a hollow cylindrical specimen, equipped with a thermal chamber controlling the temperature; (2) four cameras, in pairs, used to perform digital image correlation (DIC) analysis, in order to determine the three-dimensional strain field in the upper and lower layers and to calculate the relative displacements at the interface between different layers; (3) several displacement sensors around the specimen to control its global deformation: one pair of noncontact sensors to control displacements in the vertical direction, and another pair to control torsional displacement. The sample has a total height of 125 mm, an outer radius of 86 mm, and an inner radius of 61 mm. The small thickness of the cylindrical wall of the sample allows consideration of quasi-homogeneous strain and stress fields.

### 3. Material and Experimental Procedure

The sample tested in this study is composed of two different bituminous layers (classically used in France as base and surface layers) with an interface in between made of a tack coat (bitumen emulsion). The sample was cored from a slab produced by successive compaction of the two layers using a wheel compactor. The sample was tested at 0 °C, 10 °C, 20 °C, 30 °C, and 40 °C by applying sinusoidal torsion at 5 different frequencies, 0.01, 0.03, 0.1, 0.3, and 1Hz. The global shear strain amplitude applied was 200μm/m, while normal stress was maintained at 0 MPa during the test. The two pairs of cameras were placed on opposite sides of the sample and each camera took 50 photos per loading cycle (but 35 photos per loading cycle at the loading frequency of 1 Hz). A specific analysis [7,8] was developed at ENTPE to compute the strains in both layers and the displacement gap at the interface.

### 4. Results

The complex shear moduli $G_{\theta z}^*$ of the two bituminous mixtures (upper and lower layers) were obtained. Since Time-Temperature Superposition Principle is validated for them, master curves are plotted at 20 °C (Figure 1a for the norm, and Figure 1b for the phase angle). Corresponding shift factors are plotted in Figure 1c. The interface also shows a viscoelastic behavior as the two mixtures. Its complex shear stiffness $K_{\theta z}^*$, defined as the ratio of shear stress amplitude over the torsional displacement amplitude at the interface, could then be determined. This interface parameter exhibits a similar evolution with temperature and frequency as the viscoelastic parameter $G_{\theta z}^*$. It is then possible to plot its master curves (norm and phase angle), which are also shown in Figure 1.

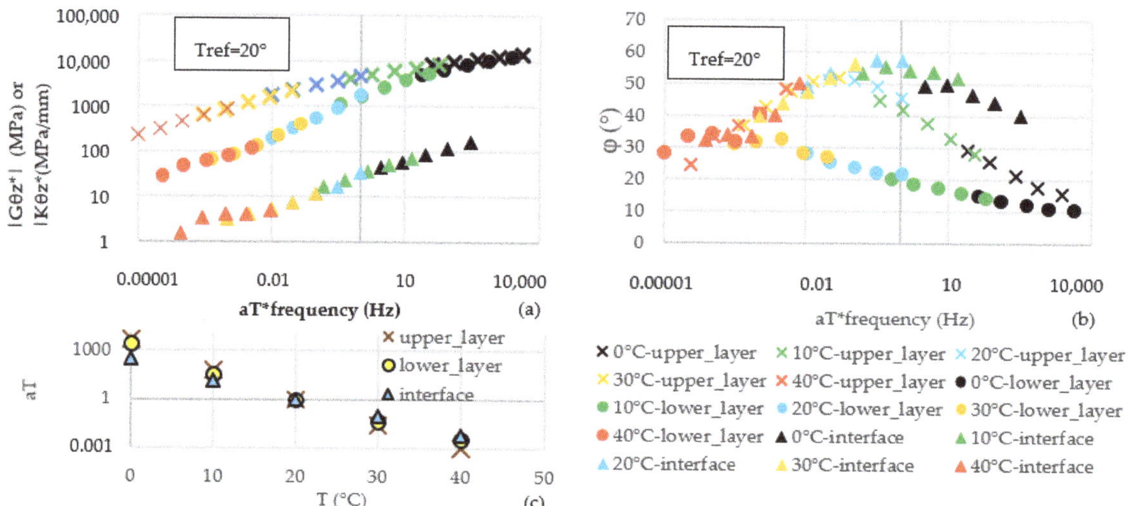


**Figure 1.** Master curves at 20 °C and shift factors of complex shear moduli of upper and lower layer and complex shear stiffness of the interface ((**a**), norm; (**b**), phase angle; (**c**), shift factors).

### 5. Discussion

The viscoelastic behavior of a multilayered sample composed of two mixtures, as well as their interface, was successfully investigated using the 2T3C apparatus. The DIC technology allows determining a displacement gap at the interface as low as 1–2 μm. These results show the potential of the device and the analysis.

**Author Contributions:** T.N.T.: Conceptualization, Methodology, Formal analysis, Investigation, Writing. S.M.: Conceptualization, Methodology, Supervision. C.S.: Conceptualization, Methodology,

Supervision. H.D.B.: Conceptualization, Methodology, Supervision. All authors have read and agreed to the published version of the manuscript.

**Funding:** This research received no external funding.

**Acknowledgments:** The authors acknowledge the company EIFFAGE for providing materials in this study.

**Conflicts of Interest:** The authors declare no conflict of interest.

# References

1. Canestrari, F.; Ferrotti, G.; Lu, X.; Millien, A.; Partl, M.N.; Petit, C.; Phelipot-Mardelé, A.; Piber, H.; Raab, C. Mechanical Testing of Interlayer Bonding in Asphalt Pavements. In *Advances in Interlaboratory Testing and Evaluation of Bituminous Materials*; Springer: Dordrecht, The Netherlands, 2013; Volume 9, pp. 303–360.
2. Canestrari, F.; Ferrotti, G.; Graziani, A. Shear failure characterization of time–temperature sensitive interfaces. In *Mechanics of Time-Dependent Materials*; Springer Science+Business Media: Dordrecht, The Netherlands, 2016; Volume 20, pp. 405–419. [CrossRef]
3. Graziani, A.; Canestrari, F.; Cardone, F.; Ferrotti, G. Time–temperature superposition principle for interlayer shear strength of bituminous pavements. *Road Mater. Pavement Des.* **2017**, *18*, 1–14. [CrossRef]
4. Ragni, D.; Ferrotti, G.; Petit, C.; Canestrari, F. Analysis of shear-torque fatigue test for bituminous pavement interlayers. *Constr. Build. Mater.* **2020**, *254*, 119309. [CrossRef]
5. Ragni, D.; Sudarsanan, N.; Canestrari, F.; Kim, Y.R. Investigation into fatigue life of interface bond between asphalt concrete layers. *Int. J. Pavement Eng.* **2021**, 1–15. [CrossRef]
6. Canestrari, F.; Attia, T.; Di Benedetto, H.; Graziani, A.; Jaskula, P.; Kim, Y.R.; Maliszewski, M.; Pais, J.C.; Petit, C.; Raab, C.; et al. Interlaboratory Test to Characterize the Cyclic Behavior of Bituminous Interlayers: An Overview of Testing Equipment and Protocols. In Proceedings of the RILEM International Symposium on Bituminous Materials, Lyon, France, 14–16 December 2020.
7. Attia, T.; di Benedetto, H.; Sauzéat, C.; Pouget, S. 3T2C HCA, a new hollow cylinder device using Digital Image Correlation to measure properties of interfaces between asphalt layers. *Constr. Build. Mater.* **2020**, *247*, 118499. [CrossRef]
8. Attia, T.; di Benedetto, H.; Sauzéat, C.; Pouget, S. Behaviour of an interface between pavement layers obtained using Digital Image Correlation. *Mater. Struct.* **2021**, *54*, 38. [CrossRef]

*engineering
proceedings*

*Abstract*

# Mechanical Properties and Structural Requirements of Recycled Aggregate Concrete for Pavements †

**Onur Ozturk** *, **Hasan Yildirim**, **Nilufer Ozyurt** and **Turan Ozturan**

Department of Civil Engineering, Bogazici University, Istanbul 34342, Turkey; hasan.yildirim@boun.edu.tr (H.Y.); nilufer.ozyurt@boun.edu.tr (N.O.); ozturan@boun.edu.tr (T.O.)
* Correspondence: onur.ozturk2@boun.edu.tr
† Presented at the 1st International Online Conference on Infrastructures, 7–9 June 2022; Available online: https://ioci2022.sciforum.net/.

**Keywords:** concrete pavements; recycled aggregates; structural requirements; recycling

**Citation:** Ozturk, O.; Yildirim, H.; Ozyurt, N.; Ozturan, T. Mechanical Properties and Structural Requirements of Recycled Aggregate Concrete for Pavements. *Eng. Proc.* **2022**, *17*, 27. https://doi.org/10.3390/engproc2022017027

Academic Editor: Piergiorgio Tataranni

Published: 2 May 2022

**Publisher's Note:** MDPI stays neutral with regard to jurisdictional claims in published maps and institutional affiliations.

In recent years, the recycling of waste materials has attracted considerable attention due to the scarcity of natural resources on earth. For instance, researchers have been working on various techniques to utilize construction and demolition wastes as substitutes for natural materials in the construction industry, which is one of the major consumers of natural resources. From this point of view, concrete roads, as one of the most frequently used infrastructural facilities [1] of the construction industry, have significant environmental impacts during their construction period and service life in different aspects. Therefore, great importance should be given to the construction of concrete roads to minimize their environmental impacts, considering their dependence on the high volume of concrete production. Using recycled materials to produce concrete roads is one of the methods implemented in this respect, as their usage provides benefits through natural resources and landfill conservation [2].

The use of recycled aggregates to reduce the environmental impact of concrete is a well-known method, and their impacts on performance have been studied in various respects to date. However, the number of studies on structural requirements of concrete pavements produced with recycled aggregates is very limited. This study aimed to investigate the structural performance of concrete pavements produced with recycled coarse aggregates as a total (100%) and partial (50%) replacement of natural coarse aggregates.

To this end, three different concrete pavement mixtures (Control, RAC-50, and RAC-100) were designed and tested for compressive strength, modulus of elasticity, flexural strength, and density. Then, material parameters obtained from the applied tests were used to determine the required thickness values for a sample pavement (based on IRC 58 [3]), and the results were compared.

According to the test results, the percent reduction compared to control mixture in average compressive strength, modulus of elasticity, flexural strength, and density values were 12.7, 7.7, 16.5, and 2.5 for RAC-50, and 18.9, 14.0, 24.9, and 4.5 for RAC-100, respectively. The test results indicate that the reduction in all the measured parameters increased with an increase in the replacement ratio of natural coarse aggregates with recycled coarse aggregates. The required concrete pavement thickness values for the control, RAC-50, and RAC-100 mixtures were determined to be 18, 20, and 23 cm, respectively (for the sample road and traffic data considered). The required thickness increased with an increase in the amount of recycled aggregate utilized (11% and 25% increase for RCA-50 and RCA-100, respectively). Additionally, concrete pavement thickness values were well-correlated with the flexural strength values obtained for the corresponding concrete mixtures.

To summarize, this study numerically presented the change in the mechanical performance of concrete due to the replacement of natural coarse aggregate with recycled aggregate and the effect of obtained performance on the thickness requirement of a sample pavement. It should be noted that the test results are dependent on the properties of recycled aggregate used in this study, and various aggregate sources may yield different results.

**Author Contributions:** Conceptualization, O.O., H.Y., N.O. and, T.O.; methodology, O.O. and H.Y., validation, O.O., H.Y., N.O and T.O.; formal analysis, O.O. and H.Y.; investigation, O.O. and H.Y.; writing—original draft preparation, O.O. and H.Y; writing—review and editing, O.O., H.Y., N.O. and T.O; supervision, N.O. and T.O. All authors have read and agreed to the published version of the manuscript.

**Funding:** This research received no external funding.

**Conflicts of Interest:** The authors declare no conflicts of interest.

# References

1. Abut, Y.; Yildirim, S.T.; Ozturk, O.; Ozyurt, N. A comparative study on the performance of RCC for pavements casted in laboratory and field. *Int. J. Pavement Eng.* **2022**, *23*, 1777–1790. [CrossRef]
2. Ozturk, O.; Yildirim, H.; Ozyurt, N.; Ozturan, T. Evaluation of mechanical properties and structural behaviour of concrete pavements produced with virgin and recycled aggregates: An experimental and numerical study. *Int. J. Pavement Eng.* **2022**. [CrossRef]
3. Indian Roads Congress. Guidelines for the design of plain jointed rigid pavements for highways (IRC 58). In Proceedings of the Indian Roads Congress, New Delhi, India, June 2015; Fourth revision. Available online: https://law.resource.org/pub/in/bis/irc/irc.gov.in.058.2015.pdf (accessed on 5 June 2022).

*Abstract*

# Reclaimed Asphalt and Alkali-Activated Slag Systems: The Effect of Metakaolin [†]

Juliana O. Costa [1,2,*], Flavio A. dos Santos [2], Augusto C. S. Bezerra [2], Paulo H. R. Borges [2], Johan Blom [1] and Wim Van den bergh [1,*]

[1] EMIB Research Group, Faculty of Applied Engineering, University of Antwerp, 2020 Antwerp, Belgium; johan.blom@uantwerpen.be

[2] Department of Civil Engineering, Federal Center for Technological Education of Minas Gerais (CEFET-MG), Belo Horizonte 30510-000, Brazil; flaviosantos@cefetmg.br (F.A.d.S.); augustobezerra@cefetmg.br (A.C.S.B.); paulo.borges@cefetmg.br (P.H.R.B.)

[*] Correspondence: juliana.oliveiracosta@uantwerpen.be (J.O.C.); wim.vandenbergh@uantwerpen.be (W.V.d.b.)

[†] Presented at the 1st International Online Conference on Infrastructures, 7–9 June 2022; Available online: https://ioci2022.sciforum.net/.

**Keywords:** recycled aggregates; geopolymers; concrete; alkali-activation; pavements

**Citation:** Costa, J.O.; dos Santos, F.A.; Bezerra, A.C.S.; Borges, P.H.R.; Blom, J.; Van den bergh, W. Reclaimed Asphalt and Alkali-Activated Slag Systems: The Effect of Metakaolin. *Eng. Proc.* **2022**, *17*, 28. https://doi.org/10.3390/engproc2022017028

Academic Editor: Emiliano Pasquini

Published: 2 May 2022

**Publisher's Note:** MDPI stays neutral with regard to jurisdictional claims in published maps and institutional affiliations.

## 1. Introduction

The demolition of old flexible pavements layers generates large quantities of reclaimed asphalt pavement (RAP). This material can be milled and reused for new pavement sections as aggregates. Alkali-activated materials (AAM) are alternative cementitious binders with high strength and chemical resistance [1]. The main problem associated with RAP in cementitious materials is the strength loss due to the porous interface [2–5]. The use of metakaolin (MK) can improve slag based AAM's properties [6,7]. The objective of this study is to investigate if the properties of RAP-AAM produced with low alkali concentration (4% $Na_2O$ and Ms = 0 and 1) can be improved with 5% MK replacement. This investigation compared isothermal calorimetry, compressive and flexural strength results for RAP-AAM produced with and without MK replacement.

## 2. Materials

The RAP-AAM was produced mixing a powdered precursor with an alkali solution. The precursors used were ground granulated blast furnace slag (GGBFS) supplied by Ecocem and MK (Caltra). The alkali solution was prepared using sodium hydroxide sodium silicate solution form VWR. Fine RAP aggregate was obtained by removing the fine fraction (<4 mm) of a locally milled flexible pavement supplied by Willemen Infra Recycling. Table 1 shows the compositions studied.

**Table 1.** Compositions (Ms = silica modulus, w = water, p = precursor, a = fine RAP aggregate).

| | Precursor | | Alkali-Solution | | | a/p |
|---|---|---|---|---|---|---|
| | GGBFS | MK | $Na_2O$ | Ms | w/p | |
| R4-0 | 100.0 g | 0 g | 4% | 0 | 0.5 | 1.5 |
| 5MK4-0 | 95.0 g | 4.5 g | 4% | 0 | 0.5 | 1.5 |
| R4-1 | 100.0 g | 0 g | 4% | 1 | 0.5 | 1.5 |
| 5MK4-1 | 95.0 g | 4.5 g | 4% | 1 | 0.5 | 1.5 |

Mortar mixing details and experimental procedures can be found elsewhere [5].

## 3. Results and Discussion

Figure 1 presents the calorimetry results for the different RAP-AAM samples studied. The sharp peak at the start of the experiment was only partially captured. The second and main peak is related to the precipitation of the reaction products [8,9]. Samples with sodium silicate (R4-1 and 5MK4-1) have a delayed second peak due to the workability retention and the reduced availability of OH- [10,11]. Replacing 5% of GGBFS with MK decreased the intensity and delayed the second peak. It also reduced the cumulative heat of the samples. The retarding effect of MK in the formation of reaction products was also observed in another research [12].

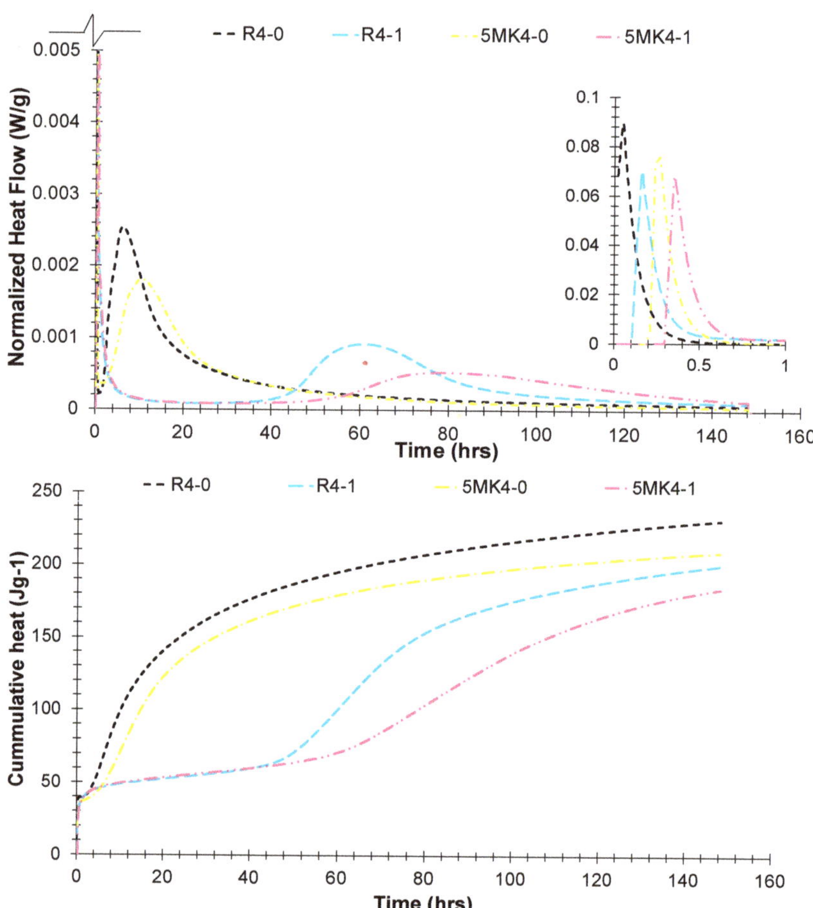



**Figure 1.** Calorimetry results of RAP-AAM.

The compressive and flexural strength of the samples is presented in Figure 2. The use of MK reduced the early strength of the samples (both compressive and flexural). At later ages, however, the use of MK caused some slight improvements in strength. This result differs from other studies [7,13] that reported a reduction in compressive strength and gains in flexural strength for sodium hydroxide alkali-activated pastes and mortars.

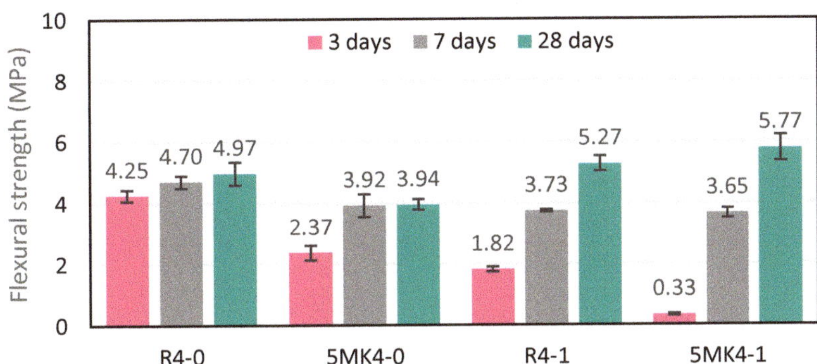

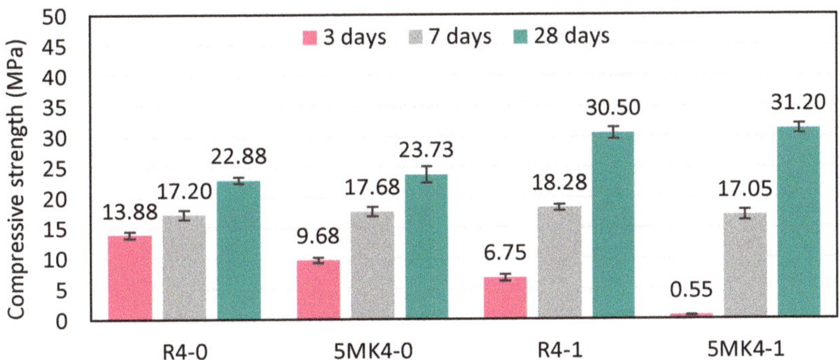

**Figure 2.** Compressive and flexural strength results for RAP-AAM.

### 4. Conclusions

The use of MK delayed the formation of main reaction products, which significantly impacted the early strength of the studied mixes. The filler effect of MK may have helped anchor RAP particles to the matrix and caused slight improvements in compressive strength observed at 28 days. This study did not see significant improvement in flexural strength as observed elsewhere [7], most likely due to the low alkali concentration used. Further studies of the benefit of MK for RAP-AAM at higher concentration of alkalis is needed.

**Author Contributions:** Conceptualization and writing—original draft preparation, J.O.C.; writing—review and editing, P.H.R.B.; supervision, J.B., W.V.d.b. and A.C.S.B.; funding acquisition, W.V.d.b. and F.A.d.S. All authors have read and agreed to the published version of the manuscript.

**Funding:** This research was financed in part by Coordenação de Aperfeiçoamento de Pessoal deNível Superior Brazil (CAPES) [Finance code 001]; and DOCPRO1 Project 42476 UAntwerp, Universiteit Antwerpen, Faculty of Applied Engineering (research group EMIB)-[Finance code 001.BOF].

**Institutional Review Board Statement:** Not applicable.



### References

1. Shi, C.; Ana, F.-J.; Palomo, A. New cements for the 21st century: The pursuit of an alternative to Portland cement. *Cem. Concr. Res.* **2011**, *41*, 750–763. [CrossRef]
2. Singh, S.; Ransinchung, G.D.; Debbarma, S.; Kumar, P. Utilization of reclaimed asphalt pavement aggregates containing waste from Sugarcane Mill for production of concrete mixes. *J. Clean. Prod.* **2018**, *174*, 42–52. [CrossRef]

3.  El Euch Ben Said, S.; El Euch Khay, S.; Loulizi, A. Experimental Investigation of PCC Incorporating RAP. *Int. J. Concr. Struct. Mater.* **2018**, *12*. [CrossRef]
4.  Brand, A.S.; Roesler, J.R. Bonding in cementitious materials with asphalt-coated particles: Part I—The interfacial transition zone. *Constr. Build. Mater.* **2017**, *130*, 171–181. [CrossRef]
5.  Costa, J.O.; Borges, P.H.R.; dos Santos, F.A.; Bezerra, A.C.S.; Blom, J.; Van den bergh, W. The Effect of Reclaimed Asphalt Pavement (RAP) Aggregates on the Reaction, Mechanical Properties and Microstructure of Alkali-Activated Slag. *CivilEng* **2021**, *2*, 794–810. [CrossRef]
6.  Bernal, S.A.; Rodríguez, E.D.; Mejia De Gutiérrez, R.; Provis, J.L.; Delvasto, S. Activation of metakaolin/slag blends using alkaline solutions based on chemically modified silica fume and rice husk ash. *Waste Biomass Valorization* **2012**, *3*, 99–108. [CrossRef]
7.  Li, Z.; Liang, X.; Chen, Y.; Ye, G. Effect of metakaolin on the autogenous shrinkage of alkali-activated slag-fly ash paste. *Constr. Build. Mater.* **2020**, *278*, 1–11. [CrossRef]
8.  Fernandez-Jimenez, A.; Puertas, F.; Arteaga, A. Determination of kinetic equations of alkaline activation of blast furnace slag by means of calorimetric data. *J. Therm. Anal. Calorim.* **1998**, *52*, 945–955. [CrossRef]
9.  Zhou, H.; Wu, X.; Xu, Z.; Tang, M. Kinetic study on hydration of alkali-activated slag. *Cem. Concr. Res.* **1993**, *23*, 1253–1258. [CrossRef]
10. Liu, S.; Li, Q.; Han, W. Effect of various alkalis on hydration properties of alkali-activated slag cements. *J. Therm. Anal. Calorim.* **2018**, *131*, 3093–3104. [CrossRef]
11. Kashani, A.; Provis, J.L.; Qiao, G.G.; Van Deventer, J.S.J. The interrelationship between surface chemistry and rheology in alkali activated slag paste. *Constr. Build. Mater.* **2014**, *65*, 583–591. [CrossRef]
12. Li, Z.; Nedeljković, M.; Chen, B.; Ye, G. Mitigating the autogenous shrinkage of alkali-activated slag by metakaolin. *Cem. Concr. Res.* **2019**, *122*, 30–41. [CrossRef]
13. Fu, B.; Cheng, Z.; Han, J.; Li, N. Understanding the role of metakaolin towards mitigating the shrinkage behavior of alkali-activated slag. *Materials* **2021**, *14*, 6962. [CrossRef] [PubMed]

*Abstract*

# A Bim Approach for the Design of a 5D Model of Industrial Warehouses in the Marine Environment †

Rassidatou Abbo [1], Okpwe Mbarga Richard [1], Lezin Seba Minsili [1,*] and Mbondo Jean Marc [2]

[1] Department of Civil Engineering, National Advanced School of Engineering of Yaounde, P.O. Box 8390 Yaounde, Cameroon; rshirine4@gmail.com (R.A.); okrichar@gmail.com (O.M.R.)

[2] Department Head, Port Asset Management, DAME, The Port Authority of Kribi, P.O. Box 203 Kribi, Cameroon; jean.mbondo@pak.cm

* Correspondence: seba.lezin@univ-yaounde1.cm

† Presented at the 1st International Online Conference on Infrastructures, 7–9 June 2022; Available online: https://ioci2022.sciforum.net/.

**Keywords:** design; warehouses; construction; BIM model; marine environment



**Citation:** Abbo, R.; Richard, O.M.; Minsili, L.S.; Marc, M.J. A Bim Approach for the Design of a 5D Model of Industrial Warehouses in the Marine Environment. *Eng. Proc.* **2022**, *17*, 29. https://doi.org/10.3390/engproc2022017029

Academic Editor: Ana Sánchez Rodríguez

Published: 2 May 2022

**Publisher's Note:** MDPI stays neutral with regard to jurisdictional claims in published maps and institutional affiliations.



BIM (building information modelling) is transforming the architecture, engineering, and construction (AEC) industry all around the world. In Sub-Saharan countries, its spread is in an earlier stage, while academics are working hand in hand with the local industry for its smooth implementation. In this context, the aim of this research is to provide an approach for designing industrial warehouses subjected to marine conditions using BIM. For this purpose, and considering our context, we make use of a methodology with seven steps:

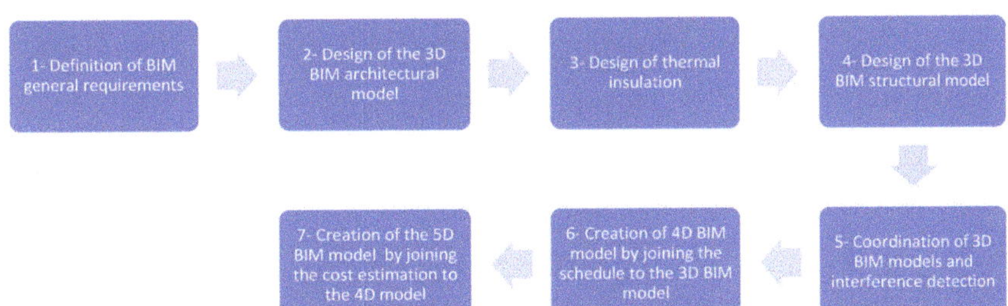


More precisely:

- Definition of BIM general requirements for this type of construction project. Here, we define the units, the language, the open standard for exchanging data, the BIM deliverables, the quality control process, and how data sharing will be performed, and adopt a Level of Development 300;
- Design of the 3D BIM architectural model. Parametric objects (footings, walls, windows, beam, column...) are used to create the model, and the software used is Autodesk Revit Architecture 2018;
- Design of thermal insulation making use of CSTB (1975), Microsoft Excel 2010, and the previous BIM architectural model;
- Design of the 3D BIM structural model using Revit 2018 Platform, Robot Structural Analysis 2018, and IFC format;
- Coordination of 3D BIM models and interference detection with the software Autodesk Navisworks Manage; all possible clashes between the different models are corrected in order to obtain a consistent 3D BIM model;

- Creation of 4D BIM model by combining the schedule (created with the software Ms. Project) of the project to each BIM objects of the 3D BIM model;
- Creation of the 5D model that gives us the cost estimation of the elements built onsite and the element to be built at every step of the project (this calculation is carried out using Navisworks manage or intelligent BIM objects).

This methodology is applied for the design of a warehouse dedicated to containing cocoa or coffee products requiring a homogeneous thermo-hydroscopic setting in a marine environment of the industrial area of the deep-sea port of Kribi (Cameroon), with a surface of 2000 m$^2$ and 11.2 m height. Preliminary results show that the proposed methodology can be easily implemented with available BIM software commonly used by engineers in Cameroon. This approach makes it possible to quickly obtain a consistent 5D BIM model of the industrial building, namely a comprehensive model which integrates data related to: architecture, structure, thermal insulation, and planning of the industrial building.

Our research studies are moving forward in order to automatically generate costs related to the project using state of art approaches related to higher Degree BIM models and based on intelligent BIM objects.

**Author Contributions:** Conceptualization, L.S.M.; methodology, O.M.R.; software, R.A.; validation, L.S.M. and M.J.M.; formal analysis, R.A.; investigation, M.J.M.; resources, L.S.M.; data curation, O.M.R.; writing—original draft preparation, R.A.; writing—review and editing, O.M.R.; visualization, R.A.; supervision, L.S.M.; project administration, L.S.M.; funding acquisition, Ongoing procedure. All authors have read and agreed to the published version of the manuscript.

**Funding:** This research received no external funding and The APC was funded by personal expenses.

**Institutional Review Board Statement:** Not applicable.

**Informed Consent Statement:** Not applicable.

**Data Availability Statement:** Not applicable.

**Conflicts of Interest:** The authors declare no conflict of interest.

*Abstract*

# Numerical Simulation of Pavement Subbase Layer Modified with Recycled Concrete Aggregates and Tire Derived Aggregates [†]

Neetu G. Kumar [1], Avishreshth Singh [2,*] and Krishna Prapoorna Biligiri [1]

1    Department of Civil and Environmental Engineering, Indian Institute of Technology, Tirupati 517 619, India; neetugkumar@iittp.ac.in (N.G.K.); bkp@iittp.ac.in (K.P.B.)
2    Indian Institute of Technology, Tirupati 517 619, India
*    Correspondence: ce18d001@iittp.ac.in
†    Presented at the 1st International Online Conference on Infrastructures, 7–9 June 2022; Available online: https://ioci2022.sciforum.net/.

**Keywords:** numerical simulation; pavement subbase; recycled concrete aggregates; tire derived aggregates; finite element analysis; sustainability

**Citation:** Kumar, N.G.; Singh, A.; Biligiri, K.P. Numerical Simulation of Pavement Subbase Layer Modified with Recycled Concrete Aggregates and Tire Derived Aggregates. *Eng. Proc.* **2022**, *17*, 30. https://doi.org/10.3390/engproc2022017030

Academic Editor: Cédric Sauzéat

Published: 2 May 2022

**Publisher's Note:** MDPI stays neutral with regard to jurisdictional claims in published maps and institutional affiliations.

The utilization of waste materials in pavement systems such as recycled concrete aggregates (RCA) and tire derived aggregates (TDA) has become a common practice in the design of surface wearing course layers. Though research is emerging on the use of RCA and TDA in the subbase layers, very limited studies are available that quantify their effect on the overall behavior of the pavement systems. Therefore, the major objective of this study was to develop a framework that could assist in quantifying pavement performance responses by using finite element method under standard wheel load of 80 kN. The information pertinent to the material characteristics of pavement systems comprising three distinct subbase layers were collected, and the designs were performed in accordance with global pavement design guidelines. The control pavement system comprised granular subbase layer, while the other two pavement designs consisted of subbase layers, which utilized RCA, and blends of RCA and TDA (RCA-TDA) as alternatives to natural aggregates. Further, axisymmetric finite element models of the three pavement systems resting over the subgrade were generated, and the stresses and strains developed in the different layers of the pavement were quantified. The test results indicated that the magnitude of vertical compressive strains for the combined RCA-TDA subbase were the highest, followed by subbase layers with RCA and natural aggregates designed separately. However, it is important to mention that the cost of 1 km long and 3.5 m wide pavement subbase with coarse granular aggregates was about 45.34% higher than the RCA subbase course and 18.74% higher than the combined RCA-TDA subbase layer. Though recycling of waste materials such as RCA and RCA-TDA resulted in slightly higher stresses and strains compared to pavement systems with virgin granular materials, the cost of construction reduced significantly along with the decreased need for extraction of virgin materials, which is certainly an approach towards low-impact development sustainable infrastructure. The framework proposed in this research may be extended further by incorporating variable traffic and different layer thicknesses or materials to ascertain the performance of a diversified set of pavements. It is envisioned that this research will not only assist in understanding the structural response of various pavement systems from a holistic design point of view, but also in promoting recycling of waste materials as applications in pavement technology from sustainability perspective, i.e., focused on waste-to-wealth and circular economy concepts.

**Author Contributions:** Conceptualization, N.G.K., A.S. and K.P.B.; methodology, N.G.K., A.S.; software, N.G.K.; validation, N.G.K.; formal analysis, N.G.K. and A.S.; investigation, N.G.K., A.S.; resources, K.P.B.; data curation, N.G.K. and A.S.; writing—original, N.G.K., A.S. and K.P.B.; draft preparation, N.G.K., A.S. and K.P.B.; writing—review, A.S. and K.P.B.; editing, A.S. and K.P.B.; visualization, N.G.K., A.S. and K.P.B.; supervision, A.S. and K.P.B.; project administration, A.S. and K.P.B.; funding acquisition, K.P.B. All authors have read and agreed to the published version of the manuscript.

**Funding:** This research received no external funding.

**Institutional Review Board Statement:** Not applicable.

**Informed Consent Statement:** Not applicable.

**Data Availability Statement:** Not applicable.

**Conflicts of Interest:** The authors declare no conflict of interest.

*engineering proceedings*

MDPI

*Abstract*

# Sustainability of Infrastructure and the Need for a Reassessment [†]

Lakshmi Thotakura [1,*], Ganesh Babu Kodeboyina [1], Deepti Avirneni [1] and Sankar Kumar Reddy Pullalacheruvu [2]

[1] Department of Civil Engineering, Mahindra University, Hyderabad 500043, India; ganesh.kodeboyina@mahindrauniversity.edu.in (G.B.K.); deepti.avirneni@mahindrauniversity.edu.in (D.A.)

[2] Department of Civil Engineering, Mahatma Gandhi Institute of Technology, Hyderabad 500075, India; pskumarreddy_civil@mgit.ac.in

[*] Correspondence: Lakshmi20pcie001@mahindrauniversity.edu.in

[†] Presented at the 1st International Online Conference on Infrastructures, 7–9 June 2022; Available online: https://ioci2022.sciforum.net/.



**Citation:** Thotakura, L.; Kodeboyina, G.B.; Avirneni, D.; Pullalacheruvu, S.K.R. Sustainability of Infrastructure and the Need for a Reassessment. *Eng. Proc.* **2022**, *17*, 31. https://doi.org/10.3390/engproc2022017031

Academic Editor: Zhenming Li

Published: 2 May 2022

**Publisher's Note:** MDPI stays neutral with regard to jurisdictional claims in published maps and institutional affiliations.



**Abstract:** The increased awareness of the effects of ecological imbalances associated with construction and industry forced several corporate and governmental bodies to look at avenues for sustainability over a broad spectrum in the 21st century. Most of these industrial and other associations, both governmental and private, started to look for the path to sustainability in a wide variety of sectors ranging from energy, urban development, corporate, agriculture, food, and even in fashion, to meet the requirements through the three known pillars of sustainability, namely environmental, societal, and economic. Coming to infrastructure, sustainability is a crucial part where the activities of design, construction, conservation of resources for future generations, could produce light-weight resilient structures having high strength and performance which improves the life span of the structure. Sustainability of infrastructure and its intricacies plays an incredible role in the assessment methodologies and the governing principles have to satisfy the requirements of three pillars of sustainability without compromising the strength and performance of the structure. The paper is an effort to present a comprehensive outline for the sustainability of resilient infrastructure, activities related to construction and prefabrication, its importance, and its assessment methodologies available presently. Policies such as minimization of construction materials, energy conservation, and use of construction and demolition waste, apart from industrial waste byproducts which, in turn, reduces the impact on environment and also minimizes the emission of $CO_2$ are advocated. It is felt that innovative, environmentally friendly, and appropriate utilization of materials based on effective research and developmental outcomes are needed. Apart from this the suitability, appropriateness, and limitations of each of the assessment methodologies for ensuring an extended lifespan in particular for the infrastructure are discussed. The aim is leaving the smallest footprint, while suggesting the possible avenues to achieve lasting structural facilities in all forms of infrastructure in future.

**Keywords:** sustainability; infrastructure; sustainability assessment; construction; environment

---

**Author Contributions:** Conceptualization, G.B.K. and D.A.; methodology, G.B.K. and L.T.; formal analysis, L.T. and D.A.; resources, L.T. and S.K.R.P.; writing—original draft preparation, G.B.K. and L.T.; writing—review and editing, D.A. and S.K.R.P.; supervision, G.B.K. and D.A. All authors have read and agreed to the published version of the manuscript.

**Funding:** Mahindra University (Internal Funds).

**Institutional Review Board Statement:** Not Applicable.

**Informed Consent Statement:** Not Applicable.

**Data Availability Statement:** Not Applicable.

**Conflicts of Interest:** No Conflict of Interest.

*engineering proceedings*

*Abstract*

# Effects of Polypropylene Macro Fibers on the Structural Requirements, Cost and Environmental Impact of Concrete Pavements [†]

**Onur Ozturk *** and **Nilufer Ozyurt**

Department of Civil Engineering, Bogazici University, 34342 Istanbul, Turkey; nilufer.ozyurt@boun.edu.tr
* Correspondence: onur.ozturk2@boun.edu.tr
† Presented at the 1st International Online Conference on Infrastructures, 7–9 June 2022; Available online: https://ioci2022.sciforum.net/.

**Keywords:** concrete pavements; polypropylene fibers; structural performance; cost; environmental impact

---

**Citation:** Ozturk, O.; Ozyurt, N. Effects of Polypropylene Macro Fibers on the Structural Requirements, Cost and Environmental Impact of Concrete Pavements. *Eng. Proc.* **2022**, *17*, 32. https://doi.org/10.3390/engproc2022017032

Academic Editor: Arkamitra Kar

Published: 2 May 2022

**Publisher's Note:** MDPI stays neutral with regard to jurisdictional claims in published maps and institutional affiliations.

With an increasing interest in environmental issues, a variety of studies have been carried out to improve the sustainability of concrete pavements [1]. Use of structural fibers in pavement applications is one of the methods proposed to reducethe carbon footprint of concrete pavements. As the fibers allow the production of concrete pavements with lower thickness and without conventional rebars [2] (which means lower use of materials), by increasing the cracking resistance, and flexural performance of concrete. However, the number of studies that numerically present the benefit that could be obtained from macro fibers is still limited. This study has been carried out to examine how the use of polypropylene (embossed, 40 mm) fibers in varying amounts (0.25–0.50–0.75–1.00%vol.) change the required thickness, cost, and environmental impact ($CO_2$ emission) of concrete pavements. Selection of polypropylene fiber among its alternatives (steel, glass, carbon, etc.) was done by considering their common usage in slab-on-ground applications, which is due to their various advantages, such as ease of handling, competitive cost, and corrosion free nature.

To achieve the aim of the study, first an experimental study was conducted to determine the mechanical performance (compressive strength, modulus of elasticity, and flexural performance) of concrete mixtures with and without fibers. Then, thickness design for a sample road was done (according to IRC 58 [3]) by using the experimentally obtained material parameters, and specified thickness values were used to determine the amount of material (aggregate, cement, water, super-plasticizer, fiber) required to produce 1 $m^2$ pavement. In the last part, by using the amount of required materials and cost/$CO_2$ emission of unit products, cost and $CO_2$ emission values were determined for each of the considered mixtures (for 1 $m^2$ pavement construction).

Based on the mechanical test results, used fibers did not considerably change the compressive strength, modulus of elasticity, and flexural strength of concrete mixtures. However, considerable improvements in the post-cracking flexural performance were obtained for the fiber reinforced concrete (FRC) mixtures depending on the amount of fiber used. Despite the increasing cost (13.9–51.3–85.5–111.5% increase for 0.25–0.50–0.75–1.00%vol., respectively), decreased thickness requirements (5.2–9.6–14.0–19.7% reduction for 0.25–0.50–0.75–1.00%vol., respectively) and $CO_2$ emissions (8.3–9.9–11.6–15.1% reduction for 0.25–0.50–0.75–1.00%vol., respectively) were found for FRC mixtures compared to the plain one. Based on the results, despite the decrease in thickness requirement and $CO_2$ emission, material cost increases with increasing polypropylene fiber amount. It is worth noting here that the presented results are valid for the fibers used in this study, and use of

---

different fiber types (with different raw materials (e.g., recycled fibers), surface properties, lengths, aspect ratios, etc.) might alter the results in varying amounts.

**Author Contributions:** Conceptualization, O.O. and N.O.; methodology, O.O. and N.O., validation, O.O. and N.O.; formal analysis, O.O.; investigation, O.O.; writing—original draft preparation, O.O.; writing—review and editing, O.O. and N.O.; supervision, N.O. All authors have read and agreed to the published version of the manuscript.

**Funding:** This research received no external funding.

**Institutional Review Board Statement:** Not applicable.

**Informed Consent Statement:** Not applicable.

**Conflicts of Interest:** The authors declare no conflict of interest.

## References

1. Ozturk, O.; Yildirim, H.; Ozyurt, N.; Ozturan, T. Evaluation of mechanical properties and structural behaviour of concrete pavements produced with virgin and recycled aggregates: An experimental and numerical study. *Int. J. Pavement Eng.* **2022**. [CrossRef]
2. Ozturk, O.; Menekse, F.; Ozyurt, N. Usage of Fiber Reinforcement for Concrete Pavement Applications and Implication of Fiber Reinforcement to the Thickness Design. In Proceedings of the International Conference on Cement-Based Materials Tailored for a Sustainable Future-(CBMT), Istanbul, Turkey, 27–29 May 2021.
3. Indian Roads Congress. *Guidelines for the Design of Plain Jointed Rigid Pavements for Highways (IRC 58)*; 4th revision; Indian Roads Congress: New Delhi, India, 2015.

*engineering
proceedings*

*Abstract*

# Checking IFC with MVD Rules in Infrastructure:
# A Case Study †

**Štefan Jaud [1,*]** and **Sergej Muhič [2]**

[1]  Jaud IT, 82272 Moorenweis, Germany
[2]  Sergej Muhič IT Consulting, 83556 Griesstätt, Germany; sergej.muhic@pm.me
*   Correspondence: stefan.jaud@outlook.com
†   Presented at the 1st International Online Conference on Infrastructures, 7–9 June 2022;
    Available online: https://ioci2022.sciforum.net/.

**Keywords:** BIM; MVD; EIR; infrastructure

check for
**updates**

**Citation:** Jaud, Š.; Muhič, S.
Checking IFC with MVD Rules in
Infrastructure: A Case Study. *Eng.
Proc.* **2022**, *17*, 33. https://doi.org/
10.3390/engproc2022017033

Academic Editor: Andrea Graziani

Published: 2 May 2022

**Publisher's Note:** MDPI stays neutral
with regard to jurisdictional claims in
published maps and institutional affiliations.



Building information modelling (BIM) is getting increasingly used in practice as a method of consistent and continuous usage of digital information in the design, construction, and operation of buildings [1]. During recent years, the infrastructure sector of the architecture, engineering, and construction (AEC) domain has been introduced to the previously established workflows, processes, and data models to focus on the building sector [2,3]. This contribution showcases a typical workflow as applied to a bridge model, i.e., the quality checking and quality assurance (QA/QC) of digital information delivered during the design phase. We present the QA/QC process, report lessons learned, and conclude with an outlook.

A very important aspect of any information flow is ensuring received data's compliance with predefined requirements (see Figure 1). In the world of BIM, the Exchange Information Requirements (EIRs) lists all necessary information to be delivered at handover, i.e., every element with its attributes, attribute types, as well as constraints to values in attributes. The information author produces a BIM execution plan (BEP) which details the EIR as applied to the project considering the software solutions employed. The model is submitted in an agreed format, e.g., a vendor-neutral non-proprietary data in an Industry Foundation Classes (IFC) format [4]. Checking rules shall be derived from the BEP and encoded using the open data format mvdXML. The rules are used for automatic model checking of the delivered data from the BIM modelling process. Identified issues shall be reported back to the modeler using the BIM collaboration format (BCF) data format.

We showcase the QA/QC process on a bridge model from Sweden. The requirements were defined before the design commenced and shared with the design firm. For example, the EIR requires the length of an edge beam *Längd (kantbalk)* to be provided for the asset management system used by the agency. The BEP foresees this information to be provided within an IFC dataset, attached with a property set to an *IfcBeam* element. The property set is named *ePset_BaTManKantbalkOccurence* and the property *K35: Längd (kantbalk)*.

The corresponding checking rule in the mvdXML is presented in Figure 2. It checks that the type of the property's value is a length measure next to the correct naming of the property set and the property. The model submitted to the stakeholder is checked against the requirement with the following result. Out of 15 beams in the delivered dataset, 13 pass and 2 fail the described check, since they do not have the specific property set attached.

The example and the checking rules are prepared in the current official IFC4 version of the standard [4]. The scope of this version is building related with limited support for the infrastructure domain. Thus, many modelling decisions in BEP are suboptimal, which can frequently and knowingly involve misusing an established concept or an IFC entity. The spatial container for the whole bridge is chosen to be *IfcBuilding* and showcases a work-around for the lack of better alternatives, whereas the railing of the bridge modelled as an *IfcRailing*, for example, demonstrates good practice. Additionally, many elements are modelled using the placeholder entity *IfcBuildingElementProxy* and classified using less than ideal concepts, e.g., properties for objects defined in this project.

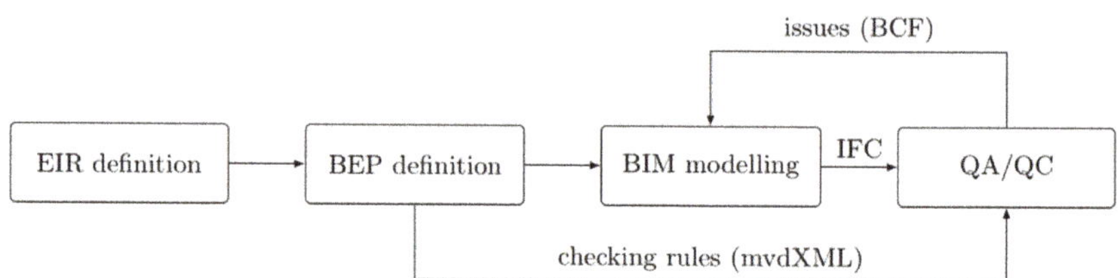


**Figure 1.** Conceptual workflow of information with QA/QC in AEC domain.

```
<TemplateRule
 ↪   Parameters="Set[Value]='ePset_BaTManKantbalkOccurence' AND
 ↪   PropertyName[Value]='K35: L&#xE4;ngd (kantbalk)' AND
 ↪   Value[Exists]=TRUE AND Value[Type]='IFCLENGTHMEASURE'"/>
```

**Figure 2.** The checking rule encoded in mvdXML.

The IFC standard was expanded over the course of the past few years to provide better support for infrastructure specifics [3]. The authors call for its fast adoption in the industry to ensure semantically rich exchanges with little-to-none work-arounds needed. This can provide a sound basis for QA/QC in the infrastructure domain of the AEC industry.

**Author Contributions:** Conceptualization, Š.J. and S.M.; methodology, S.M.; software, Š.J. and S.M.; validation, Š.J.; data curation, Š.J.; writing—original draft preparation, Š.J.; writing—review and editing, Š.J. and S.M. All authors have read and agreed to the published version of the manuscript.

**Funding:** This research received no external funding.

**Acknowledgments:** The authors thankfully acknowledge the bridge example data as well as the requirements provided by Karin Anderson from Trafikverket, Sweden.



## References

1. Borrmann, A.; König, M.; Koch, C.; Beetz, J. *Building Information Modeling—Technology Foundations and Industry Practice*; Springer: Berlin, Germany, 2018.
2. Bradley, A.; Li, H.; Lark, R.; Dunn, S. BIM for infrastructure: An overall review and constructor perspective. *Autom. Constr.* **2016**, *71*, 139–152. [CrossRef]
3. Jaud, Š.; Esser, S.; Muhič, S.; Borrmann, A. Development of IFC schema for infrastructure. In Proceedings of the 6th International Conference SiBIM: 2nd DigiPLACE RegionalConference, online, 23–24 June 2022; pp. 27–35.
4. ISO. *Industry Foundation Classes (IFC) for Data Sharing in the Construction and Facility Management Industries—Part 1: Data Schema*; Standard; International Organization for Standardization: Geneva, Switzerland, 2018.

*engineering
proceedings*

*Abstract*

# Investigating Tools for Sustainability Assessment of Road Pavements in Europe [†]

Gabriella Buttitta, Gaspare Giancontieri, Silvia Milazzo, Chiara Mignini, Patricia Hennig Osmari, Usman Ghani and Davide Lo Presti *

Department of Engineering, University of Palermo, Building 8, Viale delle Scienze, 90128 Palermo, Italy; gabriella.buttitta01@you.unipa.it (G.B.); gaspare.giancontieri@unipa.it (G.G.); silvia.milazzo02@unipa.it (S.M.); chiara.mignini@unipa.it (C.M.); patricia.hennigosmari@unipa.it (P.H.O.); usman.ghani@unipa.it (U.G.)
* Correspondence: davide.lopresti@unipa.it
† Presented at the 1st International Online Conference on Infrastructures, 7–9 June 2022; Available online: https://ioci2022.sciforum.net/.

**Keywords:** reclaimed asphalt (RA); sustainability assessment; road pavement; life cycle sustainability assessment (LCSA); sustainability rating system

**Citation:** Buttitta, G.; Giancontieri, G.; Milazzo, S.; Mignini, C.; Hennig Osmari, P.; Ghani, U.; Lo Presti, D. Investigating Tools for Sustainability Assessment of Road Pavements in Europe. *Eng. Proc.* **2022**, *17*, 34. https://doi.org/10.3390/engproc2022017034

Published: 2 May 2022

**Publisher's Note:** MDPI stays neutral with regard to jurisdictional claims in published maps and institutional affiliations.

## 1. Overview and Motivation

Sustainability assessment (SA) is a method to support decision making processes through the evaluation of system effectiveness, environmental integrity, economic valuation, and social implications [1]. SA can be carried out through the application of life-cycle-based techniques for quantitative assessment, or by performing a mainly qualitative approach via sustainability rating systems (SRS).

In the field of civil engineering, many SRS have been proposed, all based on assigning point values to actions that are determined to contribute to the overall sustainability of the project. However, only few of these systems can be applied specifically to compare road pavement technologies and/or maintenance and rehabilitation strategies. This study focuses on adapting two of these tools: GreenPave [2], developed in the US, and BE$^2$ST (Building Environmentally and Economically Sustainable Transportation–Infrastructure–Highways) [3], developed in Canada. The investigation consisted of evaluating the feasibility of increasing the amount of reclaimed asphalt (RA) in European wearing courses by carrying out a comparative analysis of eight different mixtures, containing up to 90% of RA.

## 2. Methodology, Results and Main Contribution

As anticipated above, the SA was performed using two SRS: GreenPave and BE$^2$ST. Both tools allow us to carry out an SA exercise by assigning a label to each compared alternative, from Gold to Bronze according to the final rating; however, GreenPave limits the assessment to the asphalt mixtures technology development phase, while BE$^2$ST allows us to also compare road pavement maintenance strategies. Even if there are some similarities, the scores are assigned with different criteria. In fact, if GreenPave groups the sustainability goals into four categories (Pavement technologies, Material and Resources, Energy and Atmosphere, Innovation and Design Process), BE$^2$ST judges the performance, evaluating the Life Cycle Assessment [4,5] for environmental aspects, the Life Cycle Cost Analysis for economic impacts [6], the traffic noise, the social costs, the social carbon costs and the recycling ratio. Furthermore, BE$^2$ST expresses the results as a percentage of the baseline: the label depends on the term of comparison.

In order to apply the former tool to the EU context, ECORCE M [7] was used instead of PALATE for calculating environmental indicators, while the Social Carbon Cost was assessed by considering the European average annual salary.

At first, the study provides limits and benefits of the EU-adapted SRS; then, a validation of the tools was performed by carrying out a SA of three case studies. As a result, both SRSs provide similar trends of scores when compared with hot asphalt mixtures for wearing courses with no recycled materials; however, GreenPave labels all the RA technologies as Gold or Silver, unlike conventional asphalts, which never meet the requirements for sustainability (Figure 1). On the other side, with BE$^2$ST, almost all the new mixtures achieve a label (Figure 2).

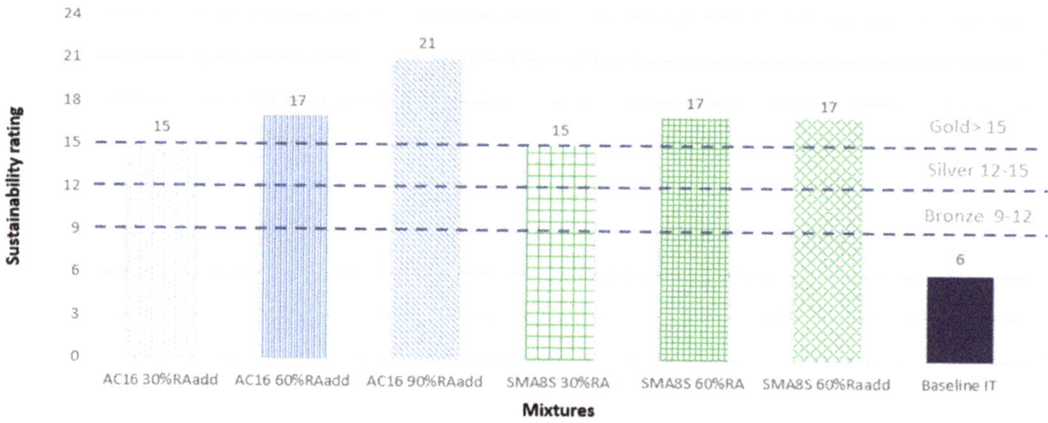


**Figure 1.** Results of the south EU case study calculated with EU-adapted GreenPave system.

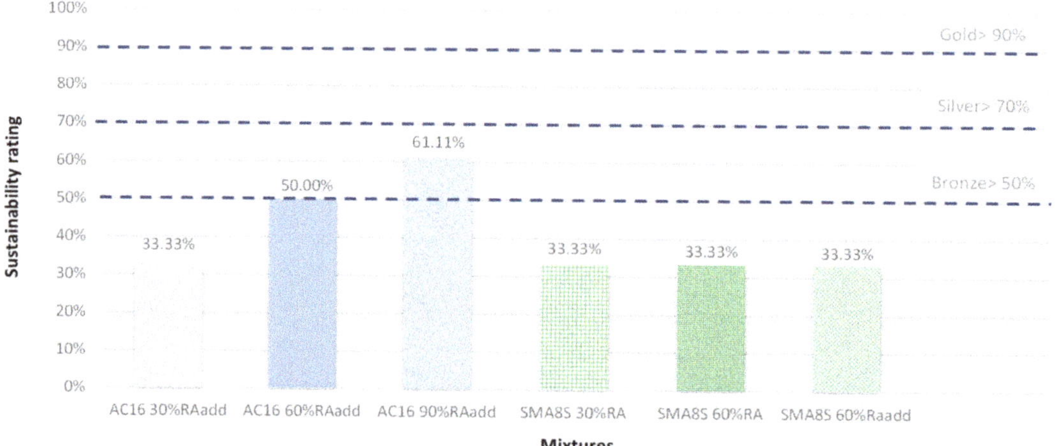


**Figure 2.** Results of the south EU case study calculated with EU-adapted BE$^2$ST system.

### 3. Conclusions and Future Works

In conclusion, it can be stated that, regardless of the SRS tools, maximizing the quantity of RA in hot mix asphalt for wearing courses, while guaranteeing the same level of durability, seems to be a more sustainable solution than not recycling at all. This is true for both a single intervention and by considering a 60-year maintenance strategy.

**Author Contributions:** Conceptualization, G.B. and D.L.P.; methodology, G.B. and D.L.P.; software, D.L.P.; validation, G.B. and D.L.P.; formal analysis, G.B. and D.L.P.; investigation, G.B. and D.L.P.; resources, D.L.P.; data curation, G.B. and D.L.P.; writing—original draft preparation, G.B., D.L.P.; writing—review and editing, G.B., D.L.P.; visualization, D.L.P., G.G., P.H.O., S.M., C.M. and U.G.;

supervision, D.L.P., G.G., P.H.O., S.M., C.M. and U.G.; project administration, D.L.P.; funding acquisition, D.L.P. All authors have read and agreed to the published version of the manuscript.

**Funding:** This research was carried out within the project "AllBack2Pave: Toward a sustainable 100% recycling of reclaimed asphalt in road pavements", funded by the Conference of Directors of Roads (CEDR) for the Call 2012 "Recycling- Road construction in a post-fossil fuel society".

**Institutional Review Board Statement:** Not applicable.

**Informed Consent Statement:** Not applicable.

**Data Availability Statement:** The results of this project are freely published within the website of the CEDR Transnational Road research Programme Call 2012 "Recycling"CEDR https://www.cedr.eu/call-2012-recycling-road-construction-post-fossil-fuel-society.

**Acknowledgments:** The research presented in this report was carried out as part of the CEDR Transnational Road research Programme Call 2012 "Recycling". The funding for the research is provided by the national road administrations of Denmark, Finland, Germany, Ireland, Netherlands and Norway.

**Conflicts of Interest:** The authors declare no conflict of interest. The funders had no role in the design of the study; in the collection, analyses, or interpretation of data; in the writing of the manuscript, or in the decision to publish the results.

## References

1. Eisenman, A.A. Sustainable Streets and Highways: An Analysis of Green Roads Rating Systems. Master's Thesis, Georgia Institute of Technology, Atlanta, GA, USA, 2012.
2. Lane, B.; Lee, S.; Bennett, B.; Chan, S. *GreenPave Reference Guide: Version 2.0*; Ontario Ministry of Transportation's Materials Engineering and Research Office: Toronto, AN, Canada, 2014.
3. Lee, J.C.; Edil, T.B.; Benson, C.H.; Tinjum, J.M. Evaluation of Variables Affecting Sustainable Highway Design with BE2ST-in-Highways. *Transp. Res. Rec. J. Transp. Res. Board Natl. Acad.* **2013**, *2233*, 178–186. [CrossRef]
4. ISO 14040:2006. Environmental Management—Life Cycle Assessment—Principles and Framework. International Organization for Standardization (ISO): Geneva, Switzerland, 2006.
5. ISO 14044:2006. Environmental Management—Life Cycle Assessment—Requirements and Guidelines. International Organization for Standardization (ISO): Geneva, Switzerland, 2006.
6. FHWA. *Life-Cycle Cost Analysis in Pavement Design—In Search of Better Investment Decisions*; Federal Highway Administration (FHWA): Washington, DC, USA, 1998.
7. Dauvergne, M.; Jullien, A.; Proust, C.; Tamagny, P.; Ventura, A.; Coelho, C.; Naullet, M.; Prual, J.M.; Fesser, A.; Tremblayt, M.; et al. *ECORCE M User's Manual*; French Institute of Science and Technology for Transport Development and Networks (IFSTTAR): Nantes, France, 2014.

*engineering proceedings*

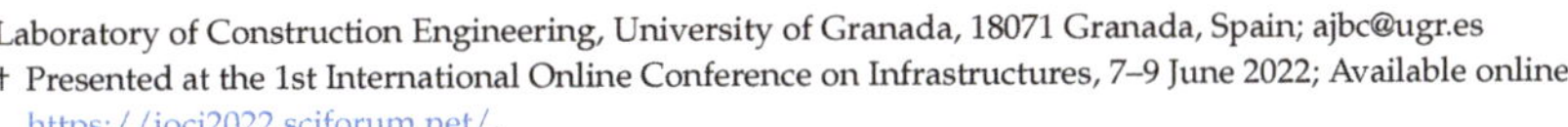

*Abstract*

# Combining Reclaimed Asphalt and Non-Petroleum-Based Binders for the Design of Sustainable Asphalt Mixtures [†]

Ana Jiménez del Barco Carrión

Laboratory of Construction Engineering, University of Granada, 18071 Granada, Spain; ajbc@ugr.es
† Presented at the 1st International Online Conference on Infrastructures, 7–9 June 2022; Available online:
https://ioci2022.sciforum.net/.

**Keywords:** biomaterials; recycling; sustainability

**Citation:** Jiménez del Barco Carrión, A. Combining Reclaimed Asphalt and Non-Petroleum-Based Binders for the Design of Sustainable Asphalt Mixtures. *Eng. Proc.* **2022**, *17*, 35. https://doi.org/10.3390/engproc2022017035

Academic Editor: Kamilla Vasconcelos

Published: 2 May 2022

**Publisher's Note:** MDPI stays neutral with regard to jurisdictional claims in published maps and institutional affiliations.

## 1. Introduction

The use of alternative materials in asphalt pavements has become a critical matter in pavement engineering due to sustainability issues. These issues include the utilisation of finite resources, the need to re-use wastes, and reduce the generation of greenhouse gas emissions. To cope with these issues, two approaches can be considered. Firstly, the use of Reclaimed Asphalt (RA) in new asphalt mixtures has become a common practice in recent decades in the asphalt industry, particularly in small amounts (<20%). However, there are still some concerns regarding the use of higher amounts (>20%) due to uncertainties in its performance. Secondly, the use of non-petroleum-based binders as alternatives to conventional bitumen is starting to gain force in this field. Recently, the combination of both approaches has been shown to be feasible and could lead to more sustainable solutions in pavement engineering. Nevertheless, more research is needed to give confidence to these innovative asphalt mixtures towards their final implementation. In this regard, the aim of this investigation is to optimise the combination of an RA source and an alternative binder made from vegetal by-products of other industries (biobinder) and target the maximum content of both materials in the asphalt mixture.

## 2. Methodology

For this purpose, two sources of reclaimed asphalt (RAs) and two types of biomaterials were characterised, namely a biobinder and a bioemulsion. The cohesion and stiffness properties of the RAs were studied by means of ITS and ITSM testing 100% RA specimens manufactured at different temperatures. With this, the degree of activation of the RA binders was estimated. On the other hand, the biobinder, bioemulsion, and the extracted binder from the RA were conventionally and rheologically characterised at the whole range of service temperatures of pavements.

The optimisation of the design of the sustainable asphalt mixtures was performed using the rheology and performance-related properties of the RA, the extracted binder from RA and the biobinder, and the different results and hypotheses on the degree of blending between both binders obtained from the RA characterisation.

## 3. Results

The results of the RAs characterisation show the potential of these methodologies to determine the degree of binder activation as an intrinsic property of RA. The characterisation of the biomaterials reveals their ability to fully replace asphalt binders in hot and cold asphalt mixtures. Finally, the optimisation of the asphalt mixture design using the rheological and performance-related properties of the individual components of the mixture and the degree of blending between binder show its key role in the design of suitable and sustainable asphalt mixtures that include alternative materials.

**Funding:** This research has received funding from the European Union's Horizon 2020 research and innovation programme under the Marie Sklodowska-Curie grant agreement No 754446 and UGR Research and Knowledge Transfer Found—Athenea3i.

**Institutional Review Board Statement:** Not applicable.

**Informed Consent Statement:** Not applicable.

**Data Availability Statement:** No data available.

**Acknowledgments:** The author would like to acknowledge RILEM TC 264 RAP TG5 for providing part of the methodology followed for RAP characterization.

**Conflicts of Interest:** The authors declare no conflict of interest.

*engineering*
*proceedings*

*Abstract*

# Alkali-Activated Materials as Alternative Binders for Structural Concrete: Opportunities and Challenges [†]

Guang Ye 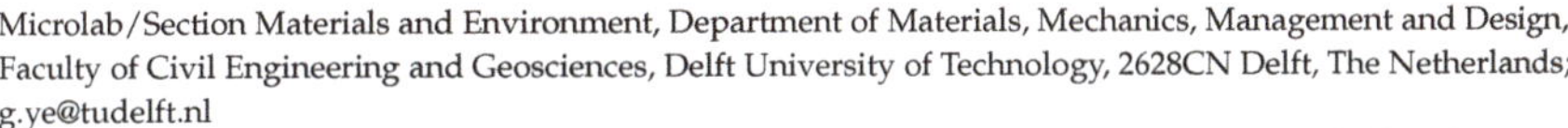

Microlab/Section Materials and Environment, Department of Materials, Mechanics, Management and Design, Faculty of Civil Engineering and Geosciences, Delft University of Technology, 2628CN Delft, The Netherlands; g.ye@tudelft.nl
† Presented at the 1st International Online Conference on Infrastructures, 7–9 June 2022; Available online: https://ioci2022.sciforum.net/.

**Keywords:** alkali-activated materials; alternative binder; materials behavior; structural application

**Citation:** Ye, G. Alkali-Activated Materials as Alternative Binders for Structural Concrete: Opportunities and Challenges. *Eng. Proc.* **2022**, *17*, 36. https://doi.org/10.3390/engproc2022017036

Academic Editor: Joaquín Martínez-Sánchez

Published: 9 May 2022

**Publisher's Note:** MDPI stays neutral with regard to jurisdictional claims in published maps and institutional affiliations.

Alkali-activated materials (AAMs, also called geopolymers) are considered as excellent alternative binders to replace Portland cement in concrete because AAMs have ce-ment clinker free binders made of industrial by-products or treated and cleaned wastes containing minerals via alkali-activation technology. AAMs have been extensively studied in the past few decades. However, industrial scale production and engineering structure applications of this type of material remain scarce. The main challenges concerning scientific and technical aspects are that: (1) qualities and chemical compositions of raw materials largely depend on the adopted processing technique and there are considerable regional differences even amongst the same kinds of materials, such as fly ash. These situations largely affect the chemical activity of raw materials and have significant influence on reaction conditions and kinetics, which consequently leads to considerable changes in the generated microstructure and entirely different behavior and performance of the material after hardening. (2) Some uncertainties regarding the long-term performances and degradation mechanisms of geopolymer systems are missing. This primary issue needs to be addressed in order to build the acceptance and confidence required for the use of AAMs in industrial applications. (3) Studies have shown that AAM concrete has different time-dependent properties (i.e., higher shrinkage and creep) compared to ordinary Portland cement concrete. This implies that when AAM concrete is used as a structural element in construction where it is restrained externally or internally, the shrinkage of geo-polymer concrete will develop a tensile stress, which might cause cracking beyond the tensile strength of the concrete.

This presentation will review recent research on these aspects and introduce some projects from materials studies to structural applications.

**Funding:** This research received no external funding.

**Institutional Review Board Statement:** The study was conducted in accordance with the Declaration of Helsinki.

**Informed Consent Statement:** The study was conducted in accordance with the Declaration of Helsinki.

**Data Availability Statement:** The data is available in the repository of the TU Delft library.

**Conflicts of Interest:** The authors declare no conflict of interest.

*engineering proceedings*

*Abstract*

# Improving Pavement Sustainability through Integrated Design, Construction, Asset Management, LCA and LCCA [†]

**John Harvey**

Department of Civil and Environmental Engineering, University of California Pavement Research Center, University of California, Davis, CA 95616, USA; jtharvey@ucdavis.edu

† Presented at the 1st International Online Conference on Infrastructures, 7–9 June 2022; Available online: https://ioci2022.sciforum.net/.

**Keywords:** pavement; asphalt; asset management; life cycle cost analysis; life cycle assessment

**Citation:** Harvey, J. Improving Pavement Sustainability through Integrated Design, Construction, Asset Management, LCA and LCCA. *Eng. Proc.* **2022**, *17*, 37. https://doi.org/10.3390/engproc2022017037

Academic Editor: Joaquín Martínez-Sánchez

Published: 12 May 2022

**Publisher's Note:** MDPI stays neutral with regard to jurisdictional claims in published maps and institutional affiliations.

## 1. Three Dimensions for Integration of Solutions to Improve Sustainability

Improving the sustainability of pavements requires action across all stages of the full life cycle of the pavement:

- Materials extraction;
- Materials processing;
- Materials transportation;
- Construction;
- Use;
- End of Life.

Proposed solutions that do not look at the complete life cycle of the pavement, and do not consider the full system (all interactions of the pavement with other systems in each stage) may result in less-than-optimal positive outcomes and create the risk of negative unintended consequences. Negative unintended consequences mean that the proposed solution may in fact achieve the opposite of sustainability goals.

The focus of most efforts concerning pavements have focused on materials, which, while also important, is only one aspect of the steps in the project delivery process where changes can be made to improve sustainability. Proposed solutions must be found in every stage of infrastructure delivery:

- Planning (if new);
- Pavement management to select project (if PMR&R);
- Conceptual design (Scoping);
- Design (PS&E);
- Construction;
- Monitor performance.

Finally, new approaches for improving pavement sustainability do not change anything until they are completely implemented, meaning that the change is embedded in policies, specifications, guidance, tools, and is part of every practitioner's everyday practice. The steps of moving from an idea to complete implementation are:

- Conceptual idea
  - Feasibility analysis using life cycle assessment (LCA) and life cycle cost analysis (LCCA) to quantify expected outcomes and cost/benefit, and further assessment of the proposed change to assess which ideas are most promising to move forward

- Research
  - ○ Reassessment as the idea is developed using LCA and LCCA to better calculate its potential for beneficial outcomes and the cost per unit of beneficial outcome
- Development
  - ○ Creating the databases, validated models, tools, policies, specifications, and training
- Implementation
  - ○ Receiving approval for implementation, making the changes in all information that is part of the project delivery process, training all users, and supporting users in their daily practice
- Feedback
  - ○ The above process concerns feedback for continuous improvement, and new concepts should be developed as the current ones are being implemented

## 2. Problems of Lack of Integration and Vision for Integrated Solutions

Recent research and development has advanced our knowledge regarding structural and material design technologies for pavements and improved methods for modeling their performance, cost and environmental impacts. However, many of these advances are not well integrated when implemented. Because of the lack of integration, advances implemented in different stages of the pavement project delivery process and network management system may not be recognized or considered in other stages. The lack of integration also results in difficulties in updating solutions in different stages of the delivery process that share common data types and models intended for the same purpose. Lack of integration presents implementation difficulties for new technologies when the implementation must be done separately for tools used in each stage of the delivery process. This presentation summarizes the overall vision and milestones reached to date for creating and implementing an integrated systems approach and continuous improvement process for the pavement enterprise in California, including structural design, materials specifications, construction specifications, network pavement asset management, life cycle cost analysis, environmental life cycle assessment, and prioritization of policies, to achieve state-wide environmental goals.

**Funding:** This research was funded by the California Department of Transportation, Partnered Pavement Research Center Agreements from 2012 through 2023 (65A0394, 65A0450, 65A0452, 65A0628, 65A0788).

**Institutional Review Board Statement:** Not applicable.

**Informed Consent Statement:** Not applicable.

**Data Availability Statement:** Presentation available at https://ioci2022.sciforum.net/#recordings (Day 1 Session 4 Part 1—Materials and Sustainability).

**Acknowledgments:** The author would like to thank colleagues at the University of California Pavement Research Center, the California Department of Transportation, and the Federal Highway Administration Sustainable Pavements Task Group.

**Conflicts of Interest:** The goals and work plans for the work presented were developed in collaboration with the project funders. Authors declare no conflict of interest.

MDPI

St. Alban-Anlage 66

4052 Basel

Switzerland

Tel. +41 61 683 77 34

Fax +41 61 302 89 18

www.mdpi.com

*Engineering Proceedings* Editorial Office

E-mail: engproc@mdpi.com

www.mdpi.com/journal/engproc

9 783036 549637